ANALYZING THE EVIDENCE

Is a Career in Forensic Science Right for You?

Evaluating Your *Fit* and Charting a Path Forward

NICOLE ZEVOTEK, M.Sc.

To the pursuit of purpose.
This book is dedicated to those looking to find
and offer their unique talents to the benefit of others.

TABLE OF CONTENTS

INTRODUCTION

Curiosity: The First Clue

My forensic science story began at about age 11, nestled in the sprawling landscape of rural Upstate New York. Our 60-acre property was a landscape of mysteries - abandoned 1940s cars half-hidden in overgrown roads, a creek dividing hillsides, and remnant stone walls of long-forgotten structures. At the base of the longest sloping hillside sat a dilapidated lean-to that my father put to use for a number of years as a hunting site and hosting his Scout troop campouts. Closer to the main and seasonal roads that bordered the property were traces of four separate structures - a house, a schoolhouse, and a farmhouse with barn. I remember the feeling of walking in the woods and coming upon any of these sites. It was a mixture of emotions - startled, tentative, curious, ...but mostly curious. Each landmark whispered stories waiting to be uncovered.

My investigative spirit manifested in what my mother considered a peculiar collection: bones, broken dishes, doll parts, leaves, fossils, rocks, fur, and feathers. My bedroom closet transformed into a makeshift Sherlock Holmes-style investigation headquarters, complete with the 1971 Encyclopedia Americana International Edition as my primary research tool.

> **SIDE NOTE:**
> I suspect even more peculiar to my mother was not the collection itself but rather the location I chose to store these collectibles (inside the house) since most were grimy, housed hibernating insects, or were still a bit fleshy.

The Innocent Investigator

My scientific curiosity wasn't without its **questionable moments.** Take my rock-smashing expedition, for instance – an unauthorized geological exploration that involved "borrowing" particularly attractive stones from a neighbor's driveway. The motivation? Pure scientific wonder: "There's gold in them thar hills!" (or at least some fascinating crystal formations).

While there's no doubt that my childhood surroundings gave way to exploration and sparked a scientific interest, it takes more than geographic locale to turn out a scientist.

> *What sparked your first moment of scientific curiosity?*
> *What experiences made you feel like a "detective"?*

These questions provide clues (data) about your interests.

> *How do childhood interests often reveal our potential career paths?*
> *How does curiosity transform into professional passion?*

These questions inform us about our shape for purpose.

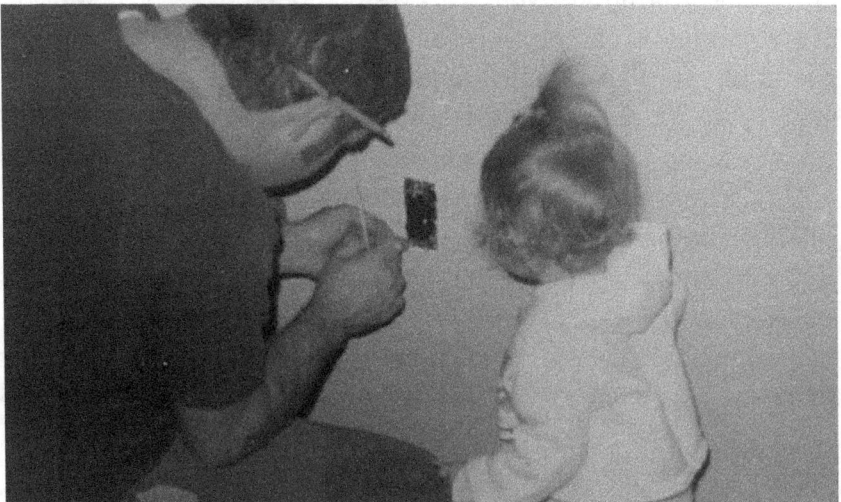

Author's first documented moment of curiosity.

The Scientific Method of Self-Discovery

Just as forensic scientists use systematic approaches to unravel mysteries, we can apply the scientific method to understanding ourselves. This book is your field guide - a comprehensive survey to help you explore, evaluate, and potentially pursue a career in forensic science.

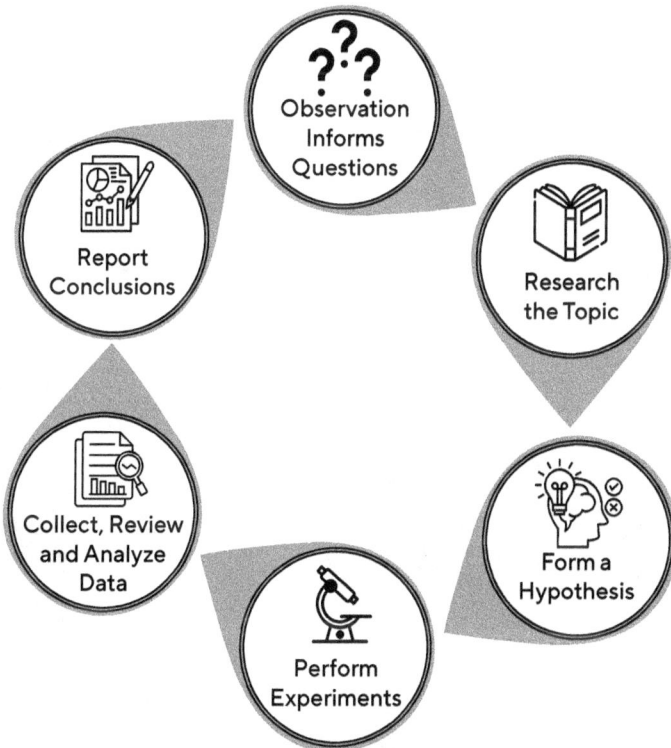

Your journey begins now. Are you ready to investigate?

1

So you want to be a Forensic Scientist?

Introducing curious minds to forensic science and the modern-day crime laboratory

If you've watched any of the popular crime scene related shows like *CSI* or *Bones* then I imagine you can probably, in an instant, picture the crime scene... it is boxed in by the signature yellow **CRIME SCENE** logo tape, triangular tags marking evidence, possibly a white tape outline of a body, and people in white hooded suits working meticulously. In most cases, the scene triggers the call for forensic science-minded expertise.

Personnel from numerous agencies may interact with the scene and the evidence at various times throughout this process. Again, if we focus solely on the physical evidence, the typical interaction at the scene involves:

Securing the scene ⇨ searching for evidence ⇨ collection of evidence ⇨ investigation of a body (if a body is present) ⇨ crime-scene reconstruction ⇨ **analysis of evidence ⇨ reporting testing results ⇨ presenting findings to members of the criminal justice system.**

Yet if we back up to about the half-way point in the above chain of events and zoom in, we find that the focus becomes about the evidence. The evidence and the information the evidence reveals about the scene, individuals, and events oftentimes provides the missing link between speculation and solvability. However not all collected evidence can or should be tested. Most crime labs do not have the resources to sustain a *test all* approach and some evidence is more relevant than other evidence.

Once evidence is secured from the scene, the totality of all the evidence collected from a scene (potentially multiple scenes) or a case must be reviewed and assessed. The assessment is referred to as *triage* where the value of the evidence determines the pathway the evidence travels, held (not probative) or evaluated (probative). Evidence typically travels as shown:

Crime Scene	Evidence Triage	Forensic Laboratory	Investigating Agency	Court Room
Physcial Evidence Collection	Probative Evidence Submitted for Testing	Evidence Tested	Action Taken Based on Results of Lab Testing	Presentation of Relevant Sciencetific Findings

At the forensic laboratory the evidence can be accessed by the various forensic science disciplines and begin to travel to separate locations (units) within the laboratory for specialized testing.

Forensic Science

In much simpler terms, forensic science is a multidisciplinary field involving the application of scientific principles and techniques to investigate crimes and legal issues. Its primary purpose is to gather, analyze, and interpret physical evidence from crime scenes to help law enforcement agencies and the criminal justice system in solving crimes and prosecuting individuals responsible for criminal activities. Forensic science plays a crucial role in the criminal justice process by providing objective and scientific support to legal investigations. Key aspects of forensic science include:

Crime Scene Processing - Recovery & Collection: Forensic scientists can begin their work at the crime scene, where they collect and document evidence such as fingerprints, blood samples, hair, fibers, firearms, and more. Proper collection and preservation of evidence are critical to maintaining its integrity for analysis.

Evidence Analysis, Instrumentation, and Technology: Once collected, the evidence is sent to specialized forensic laboratories where experts examine and analyze it. Forensic science relies heavily on advanced scientific instruments and technology, including DNA analysis equipment, chromatography, spectroscopy, and microscopy, to analyze and interpret evidence accurately.

Expert Testimony: Forensic scientists may be called upon to testify in court as expert witnesses, providing their findings and opinions to help jurors and judges understand the scientific aspects of a case. Their testimony can be crucial in the legal process.

Legal Standards: Forensic scientists must adhere to strict quality control and ethical standards to ensure the reliability and validity of their findings. They often work in accredited laboratories and follow established, validated procedures to maintain the chain of custody and prevent contamination of evidence.

Research and Development: Forensic scientists are continuously involved in research and development to improve existing techniques and develop new ones to address emerging challenges in the field.

Forensic science plays a vital role in criminal investigations, ensuring that evidence is analyzed using scientifically valid techniques and presented accurately in the courtroom to help ensure justice is served.

The Crime Laboratory

The field of forensic science is ever-expanding. Sub-disciplines and specialty areas continue to emerge as scientific techniques develop and their application within legal proceedings receive general recognition and acceptance by the forensic community. As part of the forensic community, professional organizations operate to provide guidance and standards, promote research, support improved practices, and foster collaboration to ensure the forensic science disciplines are supported and strengthened. One such organization is the American Academy of Forensic Sciences (AAFS).

AAFS is a multidisciplinary professional organization comprised of pathologists, attorneys, dentists, toxicologists, anthropologists, document examiners, digital evidence experts, psychiatrists, engineers, physicists, chemists, criminalists, educators, researchers, and others. AAFS has twelve forensic science discipline committees, but this will grow as more specializations become recognized by the forensic science community. Currently, AAFS supports the following formalized disciplines:

- anthropology
- criminalistics
- digital & multimedia science
- engineering & applied sciences
- general
- jurisprudence
- forensic nursing science
- odontology
- pathology/biology
- psychiatry & behavioral science
- questioned documents
- toxicology

Several of these disciplines are further divided. For example, the *general* section includes specialty areas such as medico-legal death investigation, crime scene investigation, forensic firearms analysis, forensic photography, forensic radiology, and forensic veterinary science.

Today, most state-of-the-art crime laboratories offer testing in toxicology, biology, drug chemistry, digital & multimedia forensics, as well as units specialized in examining trace evidence, firearms, fingerprints, and questioned documents. While there are many sub-disciplines and specialty areas of forensics, this book will cover careers in areas of study related to the core forensic science testing capabilities found in a typical crime laboratory.

Heads Up! Forensically related non-science fields such as forensic accounting, forensic art & sculpting, or those fields that require a degree in criminal justice, criminology or sociology will not be covered in this book.

FYI, I have used the term criminalist previously when referring to a professional in the field of forensics; however this title can generate confusion if you are trying to pinpoint which forensic science discipline the classification falls under. *Criminalistics* is a more generalized term often used to describe the application of scientific techniques used in the analysis, identification, and interpretation of physical evidence from a crime. A *criminalist* can possess varied skills in scientific methods and techniques to examine crime scene evidence depending on the individual's education, training, and the job duties designated by the hiring laboratory. A criminalist usually specializes in a sub-discipline of forensic science (e.g., serology & DNA analysis, fire & explosion debris analysis, trace evidence analysis) and is not a jack-of-all-trades.

Heads Up! In this book, reference will be made to the following titles to describe the forensic scientist professions: examiner, analyst, and scientist.

Soon we will focus on learning more about you, your values, abilities, and ambitions on how you can contribute to this exciting field. But for now, let's take a deeper dive into each of the forensic science roles in a modern-day crime lab to see if a particular area sparks more curiosity than another for you.

2

Forensic Professions

Explore the different career paths in the
modern crime laboratory

Crime Scene Search & Processing
Forensic Toxicology
Forensic Biology (Serology & DNA Analysis)
Forensic Chemistry
Digital & Multimedia Forensics
Trace Evidence Analysis
Firearms Analysis
Fingerprint Analysis
Questioned Document Analysis

The modern-day crime lab is a sophisticated facility equipped with advanced technologies and specialized personnel dedicated to analyzing and processing evidence related to criminal investigations.

No matter the discipline, a forensic science professional plays a crucial role in criminal investigations by providing scientific evidence that can be used to link suspects to crime scenes or exclude innocent individuals from suspicion. A forensic laboratory must have a system in place to guarantee quality and reliability of analysis (**quality assurance program**). The laboratory's policies, facility, personnel, etc., must adhere to specific standards (e.g., Federal Bureau of Investigation Quality Assurance Standards for Forensic DNA Testing Laboratories) and be approved by an appropriate authorizing organization (**accreditation process**) to permit work.

A routine part of the scientist's work involves activities established to maintain the integrity of the evidence for its future use by law enforcement and for presentation in a court of law. These activities include **accreditation**, **quality control**, **evidence handling** & **triage**, **examination/analysis**, **report writing**, **technical review**, and **testimony**. You will learn about quality assurance and accreditation in forensics in upcoming chapters.

Prior to performing any examinations on items of crime scene evidence, the forensic scientist must be 'qualified':

✓ trained in each area of testing (completion of training requirements)
✓ qualified (successful completion of competency testing)
✓ proficient (passing regular testing for proficiency in each qualified discipline)

Aside from, and often prior to, examining and testing crime scene evidence you can expect a typical day to involve any number of the following tasks:

- Cleaning laboratory benches and shared areas
- Preparing solutions and stocking supplies
- Retrieving and storing evidence
- Documenting physical observations of an item of evidence
- Collecting evidence from an item and preserving it for testing
- Preparing samples for testing
- Running an instrument
- Reviewing instrument data
- Typing reports

- Performing a technical review of another scientist's work
- Participating in a pre-trial meeting
- Providing testimony
- Receiving or providing training
- Pursuing continuing education
- Reading scientific literature or presenting a summary on the content of a scientific journal
- Participating in a proficiency test (annually or every six months, depending on the discipline)
- Performing a protocol review
- Participating in a self-assessment (audit of procedures/case reviews)
- Attending a staff meeting

Now that you have an idea of the foundational activities of a forensic laboratory professional, let's take a closer look at each of the individual professions and their specific job duties. This chapter provides an overview of the following forensic science professions: crime scene examiner, forensic toxicologist, forensic serologist & DNA analyst (forensic biology), forensic chemist, computer crimes analyst (digital & multimedia forensics), trace evidence examiner, firearms examiner, latent print examiner, and questioned document examiner.

Crime Scene Search & Processing

Crime scene personnel can operate as members of a crime laboratory system or as part of a law enforcement agency's crime scene investigation unit. The specific duties and function of each member of a crime scene unit can vary depending upon the agency type (e.g., city police, county sheriff, crime lab) and organizational structure (e.g., supervisor, investigator, officer, technician, etc.). Crime scene examiners (CSE) may go by many names, including

crime scene investigator, crime scene technician, forensic examiner, crime scene analyst, and so on. In this section we focus on the role of the crime scene search and processor and will refer to this position as the *examiner* or *CSE*.

The CSE assists in the investigation by responding to the scene of the crime, analyzing the area, searching for, locating, preserving, and collecting evidence to be submitted to the crime laboratory for testing. Specific duties may include photography, latent print processing & collection, possibly specialized collection of trace evidence and evidence for serology and/or DNA testing (swabs for DNA of handled items, swabs of body fluids, etc.).

The CSE can be called on to respond to a scene in any case category but is primarily sought to respond to major crime cases often violent in nature and/or consisting of multiple victims and/or scenes. The CSE responds to the scene once it has been secured, typically by a Patrol Officer.

Case Scenario 1: Minor Crime Scene – Responds to the scene to identify and collect evidence. Photographs and processes the scene, including latent print processing and collection of DNA swabs, as well as other relevant physical evidence.

Case Scenario 2: Digital Crime – Responds to the scene to collect videos from surveillance equipment, locates, downloads, and saves all body and in-car camera video. In this scenario, one city police department divides duties associated with videos and equipment as follows: items are collected by the crime scene unit and submitted to property control. The CSE will review the videos on scene to identify relevant areas for processing and evidence collection. Detectives will further analyze the surveillance videos. A separate, specialized unit handles all Officer Body Worn Camera (BWC) footage.

Case Scenario 3: Major Crime Scene – Responds to the crime scene to locate, preserve, and collect evidence. The scene has been secured by the Patrol Officer and may be actively being worked on by investigators and/or medical personnel upon arrival. The CSE processes the victim and the scene including taking photographs or videos, sketching or diagraming the scene, documenting measurements, collecting blood and DNA swabs, performing latent print processing and collection, etc. The CSE may respond to the medical examiner's office to collect additional evidence items from the victim's body during autopsy.

Mock crime scene.

Examples of some of the more routine sample types collected are weapons and ammunition type evidence (e.g., discharged cartridge cases, projectile fragments, copper jacketing, live ammunition, spent bullets), drugs, hair, fibers, shoe prints, fingerprints, cigarette butts, and swabs of touch DNA and body fluids on surfaces. In burglary cases the CSE may collect evidence types such as a crowbar, screw drivers, and hammers while sexual assault cases typically involve collection of a sexual assault kit at a medical facility as well as clothing and related bedding items.

The crime scene technician relies on cameras to capture photos and videos of the scene. Advanced photography techniques such as photographing a patent- or latent-type print that would be unsuccessfully lifted allows for the photograph to be used for comparison purposes. 3D scanners can be used to digitally capture and measure scenes while drones can capture aerial photographs and data to create 3D models of crime scenes. Alternative light sources and chemical screening reagents may be used to aid in illuminating body fluids and impressions at a scene for subsequent collection. These chemical reactions can also be photographed to preserve the testing result information.

Other duties for which a CSE is often involved relate to their prior processing of a scene:

Case Preparation & Reports - preparing evidence for criminal trials through detailed reports and diagrams to memorialize crime scenes as they have been found. Itemizing each item found in their reports by locations, at times with measurements.

Court - attending depositions in preparation for criminal prosecution. Appearing in court under subpoena and providing sworn testimony in court for criminal prosecution. Bringing evidence to trial to enter it into the criminal case. Testifying to the chain of custody for evidence.

BTW

A crime scene investigator (CSI) interacts with the CSE through scene containment, examination and search of the scene, recognition of types of evidence and any patterned evidence as it applies to scene interpretation, analysis, and reconstruction. The CSI is often trained in the legal/crime scene interface (e.g., search warrants), history and theory of crime scene investigations, shooting reconstruction and distance determination, nature and properties of evidence types and patterned evidence (including bloodstain pattern analysis and interpretation), crime scene safety, documentation, contamination prevention (e.g., DNA), and preservation. Keep in mind, depending on the agency/organization, the duties of the CSI and the CSE can be rolled into one position.

Forensic Toxicology

Forensic Toxicology is the study of the effect of substances (drugs, toxins, chemicals, etc.) on a person. There are two categories of testing: antemortem (subject is alive) and postmortem (subject

is deceased) testing. The most common areas of testing in this field are for human performance (such as impaired driving), workplace drug testing, criminal investigations (such as drug facilitated crimes that may include sexual assaults or poisonings), and cause of death investigations.

The types of specimens used for antemortem testing are typically blood, urine, and hair analysis. Postmortem specimens typically use blood, urine, and organs. The toxicologist is asked to examine the specimens for substances such as poisons, alcohol, and drugs and determine what their affect may be on an individual. A report of findings will be generated and used to determine a cause of death or used to aid in a criminal investigation to determine if a criminal charge is appropriate.

A laboratory can perform toxicology testing for non-criminal cases such as employment drug screening. This type of testing is often done as part of the application process for government or law enforcement agencies. Hair or blood is often collected to determine if any drugs may be present. This type of investigation is to determine if an individual uses any common drugs of abuse. The forensic toxicology unit can assist in many different types of criminal cases such as operating vehicles while intoxicated, death investigations, and non-traffic related crimes such as probation/parole violations, drug facilitated sexual assault, domestic abuse, or weapons charges. A few common scenarios are:

Case Scenario 1: Crime related to Drugs & Alcohol - An individual is pulled over for suspected driving under the influence. The toxicologist is asked to test biological samples (blood and urine) taken from the individual for the presence of alcohol and/ or drugs.

Blood and Urine Evidence Kit collected for toxicology testing. (courtesy of A. Kogelschatz)

Case Scenario 2: Postmortem Investigation – A deceased individual is found next to drug paraphernalia. The toxicologist is asked to assess biological samples from the deceased to assist in establishing the cause or circumstances of death.

Case Scenario 3: Drug facilitated crime – A person wakes up in an unknown house after being out at a bar and doesn't remember how they got there. A urine sample is collected and submitted for testing.

Toxicology testing can be performed to determine if a compound is present (qualitative analysis) and what the amount of that compound is (quantitative analysis). For forensic analysis, testing begins with a screening technique to determine a class of compounds (cocaine/metabolites, cannabinoids, methamphetamine, etc.). The type of instrumentation can be gas chromatography or immunoassay testing. A more specific test is performed for any classes of drugs that screen positive to determine the specific compound present and in what amount. This type of testing is commonly performed using headspace gas chromatography (alcohol testing), gas chromatography mass spectrometry (GC/MS), and liquid chromatography mass spectrometry (LC/MS).

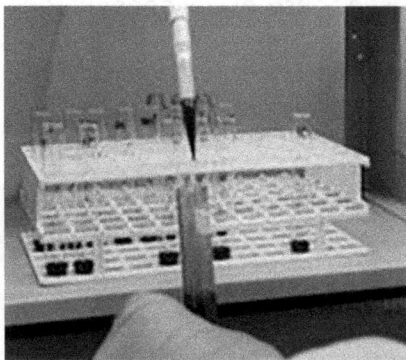

Sample Preparation – Drug Extraction method: A portion of blood from an evidence blood tube is removed to prepare it for a drug extraction method. (courtesy of A. Kogelschatz)

Forensic toxicology unit services may be requested as part of a multi-discipline case analysis. For example, in a suspicious death or potential overdose case, often drug paraphernalia can be found near the deceased individual. Evidence submitted to the crime laboratory would include a vial of blood from the victim, a syringe, and a bag of white powdery substance and spoon collected near the vic-

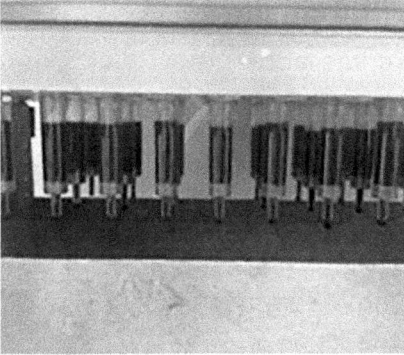

Sample Loading – Drug confirmation method: Samples are loaded on cartridges during a solid phase extraction for a drug confirmation method. The sample interacts with the cartridge material, any drugs will be retained while other materials (i.e., biological matrix, proteins, etc.) will be washed away.
(courtesy of A. Kogelschatz)

tim's body. The testing request form specifies an inquiry for toxicology and drug analysis. The forensic toxicologist will analyze the blood for the presence of controlled substances to assist in determining manner of death. The spoon and bag containing a white powdery substance will be sent to forensic drug chemistry for identification. Both of these reports would aid in determining the cause of death. Additionally, a request to swab the syringe handle and test it for DNA could be probative to the investigation if it provides information as to the handler of the syringe. This could lead investigators to other individuals who may have been present at the time of the event or clues indicating criminal behavior.

Another example of the forensic toxicology unit services being requested as part of a multi-discipline case analysis is a vehicular crash where the driver is suspected of driving under the influence case. The air bags deployed and both occupants of the vehicle denied being the driver of the vehicle. In this type of investigation, DNA analysis would be requested from the driver-side air bags to link to the individual driving the vehicle at the time of the crash. Blood would also be collected and tested for the presence of alcohol and/or drugs.

While examples of multi-disciplinary casework highlight the level of coordination of the individual crime laboratory units, it is also noteworthy to mention that a single piece of evidence can often be the source of more than one bit of information relevant to the details of the case. You will find more examples of this level of coordination in the explanations of each forensic discipline to follow.

Forensic Biology (Serology & DNA analysis)

The forensic biology unit within the crime laboratory specializes in the examination of crime scene evidence to locate and identify biological material (serology) and perform DNA testing. Identifying the type of biological material and its source through subsequent DNA testing assists the investigator in identifying victims, suspects, and placing individuals at the scene of a crime.

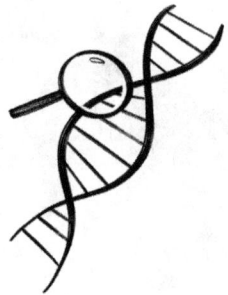

Serology

Serology involves the identification of biological material and is routinely the first step in determining how a piece of evidence might be relevant to the case.

The **forensic serologist** employs a visual exam, an additional exam assisted by fluorescent light source, chemical tests, and microscopy to screen evidence for the presence and identification of biological fluid located on an item.

The type of biological fluid recovered (i.e., semen, saliva, blood) in conjunction with the nature of the crime (i.e., sexual assault, homicide, burglary) is important in determining how relevant it might be to the case. The serologist must take this into account in determining which biological material to screen for, collect and preserve.

Clothing from a sexual assault is normally screened for blood, saliva and semen. Typically the forensic serologist screens evidence such as clothing or bedding in a prescriptive manner:

☛ **Physical Examination**: Documents physical appearance of evidence and its associated packaging. Performs preliminary visual screen for biological material and trace evidence. Determines steps of analysis (e.g., to prevent loss of hairs/fibers, trace collection may occur before chemical screening).

☛ **Trace Evidence Collection**: Plucks or uses a tape roll method to secure any hairs, fibers, or debris from fabric-type evidence such as clothing and bedding.

☛ **Crime Scope Examination**: Uses an alternative light source to illuminate and locate biological stains that may be camouflaged in a visual examination.

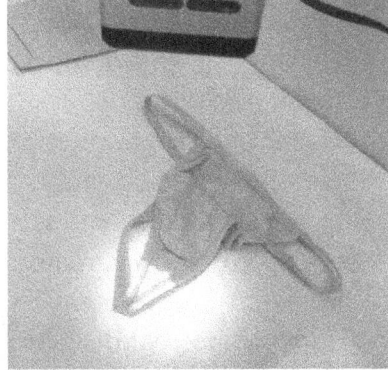

Illuminating a semen stain with the use of a CrimeScope.

☛ **Identification of Body Fluids – Chemical screening**: Employs presumptive chemical screening tests to help locate and identify stains like blood, semen, and saliva to provide information as to the type of stain to aid in discerning the nature of a crime.

☛ **Microscopic Examination**: Interprets the results of chemical screening tests. Pursues secondary tests to confirm preliminary test results when applicable (e.g., microscopic exam for spermatozoa from a suspected seminal fluid stain). Below is an example of a microscopic examination of a cutting from a stained area on the crotch of a pair of underwear. The image

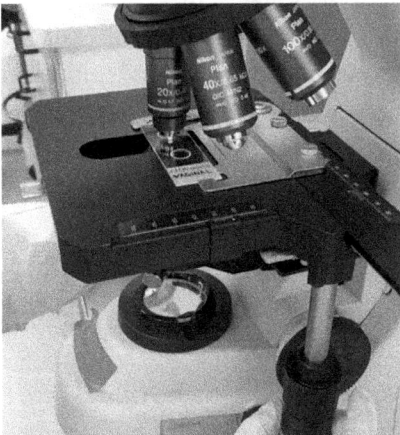

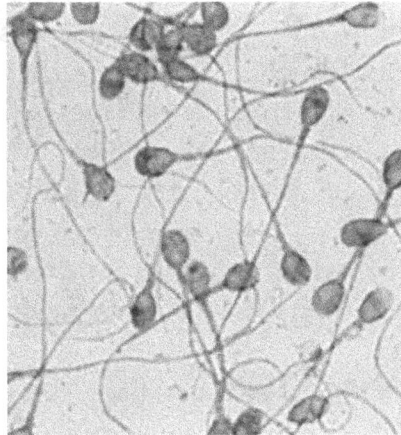

Microscopic exam on sample from a semen-stained item (left) and spermatozoa stained with Christmas Tree staining method (right).

above is a picture of spermatozoa that have been treated with Christmas Tree staining technique to enhance their visibility under a microscope.

☞ **Collection & Preservation of Biological Material**: Evaluates type, number, size, and appearance of stains present on an item of evidence in order to select probative locations to collect and preserve for downstream analysis. Collects biological material from non-stained areas on evidence items for potential DNA from contact/handling (e.g., grip/trigger of a gun). Preserves biological material such as stains, swabs, and hairs for future DNA testing.

DNA Analysis

The serological examination results are reviewed to select evidence items for further testing. Depending on the laboratory work-flow or policy, this assessment can be performed by a case manager, supervisor, or DNA analyst, or another trained individual. The **DNA analyst** will review the serologist's testing results to determine the type of testing to pursue. Laboratories validate procedures for the type of methods and instrumentation utilized in order to optimize DNA extraction and downstream processes and minimize loss or destruction of evidence. DNA analysis is the process of releasing the DNA from the cells present in the biological material (skin cells, stain, or fluid) and copying the forensically relevant locations of the genetic material to develop a DNA profile.

The DNA analyst uses a combination of chemistry and instrumentation to extract DNA and generate a DNA profile. Testing involves the following stages:

☞ **Sample Preparation:** Prepares stains, swabs, hairs, bones, etc., by removing a portion of the biological material into a tube for DNA testing.

☞ **Extraction, Quantitation, Amplification**: Uses chemicals to extract the DNA from the cells in the samples and removes non-DNA related cell material; determines quantity and quality of extracted DNA and uses Polymerase Chain Reaction (PCR) to copy forensically relevant regions of the DNA to obtain sufficient amount for analysis.

☞ **Electrophoresis**: Separates segments of genetic material from one another in a biological evidence sample using a technique called capillary electrophoresis. This method takes advantage of the charged properties of DNA and separates them by size as they travel through a charged sieve-type liquid contained within a thin capillary. The longer fragments take more time to snake through the sieve than the shorter fragments.

☞ **DNA Profiling**: Analyzes the targeted DNA fragments present in the biological material to compile into a DNA profile. Each peak represented below in the electropherogram (EPG) of the DNA profile is further named by a numerical value (8, 9, 10, etc.) which represents the number of short tandem repeats (STR) in the segment of DNA traveling through the capillary. This number serves as the basis for differentiation between individuals and is referred to as an allele.

For example, a region of DNA contains '5' short tandem repeats (STR). The peak on the EPG would be labeled '5' referring to the 5 allele. (The A, G, C, and T refer to components of DNA called nucleotides; 'A' = adenine, 'G' = guanine, 'C' = cytosine, 'T' = thymine.)

AGTC-**CTAG-CTAG-CTAG-CTAG-CTAG**-AGTC-

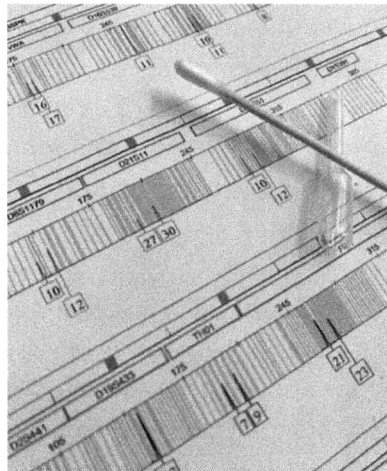

EPG of a DNA profile from a buccal swab collected from the inside cheek area of the mouth.

- ☛ **Profile Comparison**: Compares suitable DNA profiles obtained from evidence to DNA profiles from known individuals (suspect/victim – usually collected by swabbing the inside the mouth (cheek area) with a buccal swab).
- ☛ **Database Eligibility**: Recognizes forensic DNA profiles suitable for DNA database entry and submits for review of eligibility.
- ☛ **Documentation and Reporting**: Maintains detailed records of all analyses performed, drafts reports, and documents findings for legal purposes.
- ☛ **Court Testimony**: Provides expert testimony in court to explain the results of the forensic analyses.

The forensic biology unit examines evidence from many types of cases such as homicides, sexual assaults, property crimes, to possession of weapons and drugs. The most commonly submitted items in these cases are clothing and swabs of suspected blood, semen, saliva, or skin cells.

Case Scenario 1: Property Crime – A homeowner reported that an individual broke into his residence and stole a television and computer. The crime scene unit responded to the scene and swabbed the windowsill (point of entry) and the remaining TV cables and outlet where the computer had been plugged into the wall. Swabs were submitted to the lab for touch DNA testing to help identify the perpetrator.

Case Scenario 2: Sexual Assault – A young female attended an off-campus party where she believes someone may have altered her drink and sexually assaulted her after she passed out. Responding officers transported the victim to the local hospital for collection of a sexual assault evidence kit. The kit is to be tested for the presence of biological material with positive samples proceeding to DNA testing. Any forensic DNA profile recovered from the kit evidence (i.e., does not match the victim's DNA profile and any consensual sexual partner's DNA), if suitable (meets all eligibility requirements), can be entered into the Combined DNA Index System database (CODIS). Forensic DNA profiles are then searched against DNA profiles from other crime scenes and convicted offenders. If a hit occurs, the laboratory will notify the investigator of the lead.

> ### BTW
> The sexual assault evidence collection kit is an evidence type that has been specifically designed for submission to the forensic biology unit. The kit contains swabs that are collected in sets from each orifice (vaginal/penile, anal, oral) and body areas (breasts, neck, genital area) potentially involved in sexual contact. The sexual assault evidence kit is routinely submitted to the forensic biology unit for testing.

Because biological material is often recovered from the majority of crime scenes, evidence submitted to the latent print and firearms units are also sent to the forensic biology unit for testing. These crime laboratory units coordinate with each other to determine the best order of examinations.

The future of DNA analysis is ever evolving. Three techniques are being developed and implemented for use in crime laboratories and/or law enforcement agencies:

Rapid DNA – Rapid DNA instruments are being placed in booking stations for rapid development of DNA profiles from arrested individuals (DNA reference profile). The DNA reference profile from the arrestee is subsequently searched against forensic DNA samples from unsolved crimes to allow law enforcement agencies to more quickly determine (within a couple of hours vs. days) if the arrested individual was involved in a crime. Note: Keep your eye out for future use on forensic samples, including profile upload into CODIS. Police agencies using this technology for evidence items must be associated with an accredited laboratory.

Next generation sequencing (NGS) – NGS is a DNA sequencing technique that allows for millions of fragments of DNA to be sequenced simultaneously and is often used to evaluate genetic mutations down to a single nucleotide. In forensics, this technology is being used to target a selection of genetic locations with high single nucleotide variation (i.e., single nucleotide polymorphism (SNP)) that provides either bio-geographical ancestry data or data attributable to physical features such as eye color. This information makes SNPs an excellent investigative tool by providing a mechanism for discriminating one sample from another, or in this case, one individual from another.

Forensic Investigative Genetic Genealogy (FIGG) – FIGG uses forensic DNA profiles (these are STR DNA profiles developed from crime scene evidence) that are unknown (not attributed to an individual) and utilizes next generation sequencing technology to develop a different type of DNA profile (SNP DNA profile) by typing "identity SNPs" – the same type of SNP profile used for direct-to-consumer genetic testing utilized in family-matching databases. Law enforcement can subsequently load the forensic DNA SNP profile into openly accessible genealogy databases to search for genetically related DNA profiles. The resulting list of potentially related DNA profiles is further analyzed by a forensic genetic genealogist to determine if there is indeed a relationship. Case specific details such as geographical location and approximate age of the perpetrator can be compared to that of a potential relative in the database to assist investigators in narrowing down the search for a particular suspect.

Forensic Chemistry
(Also Known As: Controlled Substances/Drug Chemistry)

The forensic chemistry unit within the crime laboratory specializes in the examination of evidence for the presence of controlled substances. The federal government and each

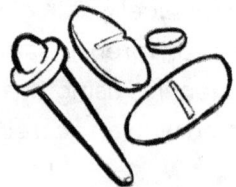

state have their own statutes regarding which substances and commonly seen drugs are *controlled*. A drug or compound that may be abused or cause addiction can be *controlled* in the way the substance is made, used, handled, stored, and distributed. Controlled substance categories include stimulants, depressants, opioids, hallucinogens, and anabolic steroids.

Instrumentation and chemical tests are used to detect the presence or absence of a controlled substance. Preliminary tests can involve a chemical spot test and/or microscopy (plant material) and are used to screen a sample of the evidence. Samples of the suspected controlled substances are then analyzed on instruments such as thin layer chromatography, gas chromatography/ mass spectrometry, infrared spectroscopy, and as chromatography/ infrared spectroscopy, then subsequently compared against known drug standards.

Forensic scientists in the forensic chemistry unit support investigations of cases involving suspected drugs. These cases can involve substances confiscated at prisons, found on an individual, found in the mail, collected through search warrants, or seized from clandestine drug laboratories.

Case Scenario 1: Crime related to Possession of a Controlled Substance – An individual is suspected of driving under the influence. The patrol car pulls the vehicle over. There is a baggy of white powder on the passenger's seat along with a used syringe. The forensic scientist is asked to test the residue from the syringe

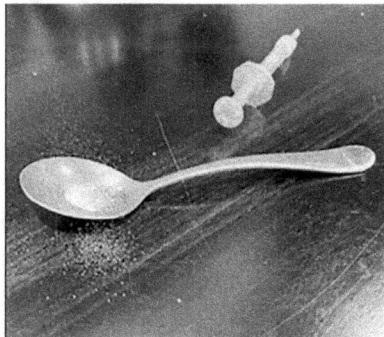
Drug paraphernalia.

and the powdered substance for suspected controlled substances.

Case Scenario 2: Crime related to Intent to Manufacture a Controlled Substance – Authorities have sufficient reason to believe that occupants of a residence are growing marijuana on their property and selling it. Plants are confiscated from the property and a single plant is submitted to the crime lab for analysis.

Examples of some of the more routine sample types are drug substances in different forms such as pills, powder, liquid, residue, or plant material.

BTW

Evidence of this nature is weighed when received by the crime laboratory in and before and after analysis by the forensic chemistry unit. This is important because the type of substance as well as the quantity is relevant in determining whether the act of possession or use of a substance is criminal in nature.

Controlled substance evidence is often contained within a small plastic or paper bag or bindle. There are scenarios where the submitting law enforcement agency would like the crime lab to determine the type of substance and attempt to identify the handler. This request for multi-discipline analysis

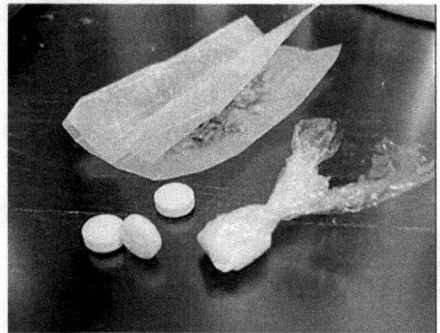

Common evidence sample types: organic material, pill, and powder.

involves the work of the forensic drug chemist and the DNA analyst. The DNA analyst will be asked to swab the bag/bindle for skin cells. This swabbing must occur prior to drug chemistry performing its tests in order to preserve the cellular material on the packaging. In some cases the drug chemist can swab the bag and preserve it for the DNA analyst to test at a later date or coordinate with a DNA analyst to collect the swabs when the evidence item is in his/her custody.

Digital & Multimedia Forensics

Electronic data is a component of everyday life and is a common source of evidence in nearly all criminal activity. Digital and multimedia forensics focus on retrieving, storing, and analyzing data from phone, computer, video, and audio sources to assist in criminal investigations by establishing facts and providing support for an investigation. Digital and multimedia forensic (DMF) services can be provided through a crime laboratory or a law enforcement agency and, depending upon the organization, the forensic examiner can hold the role of both examiner and investigator (civilian or sworn). For the purpose of describing the type of work performed in this area of forensic science, this segment will refer to this position as the Digital and Multimedia Forensic Examiner (DMFE).

The primary function of the DMF unit is to examine all types of digital evidence relating to criminal activity. By examining digital evidence, the examiner can gather identifying information that can lead to the detection of potential suspects. The DMF unit responds to calls to evaluate, collect, and process digital evidence for detectives, officers, and other agencies. The DMFE may respond to a scene to collect evidence or examine digital evidence brought to the DMF laboratory.

An important aspect of forensically examining digital evidence involves utilizing specialized hardware and software programs while ensuring the integrity of the evidence. For mobile devices this involves removing the ability of the cell phone to access the internet or wireless signals using a Faraday box - effectively securing the device in the same state as it was when seized, thereby maintaining the digital evidence in as close to its original state as possible. This preserves the data that is already stored on the device and prohibits the mobile device from receiving new data (i.e., text messages, calls received, etc.). When it comes to gaining legal access to the data, a search warrant is often necessary. A search warrant can encompass all the data stored on the device at the time of execution of the warrant; however, if new data were to be downloaded to the device after the warrant execution, access to this information would not be permitted by the warrant.

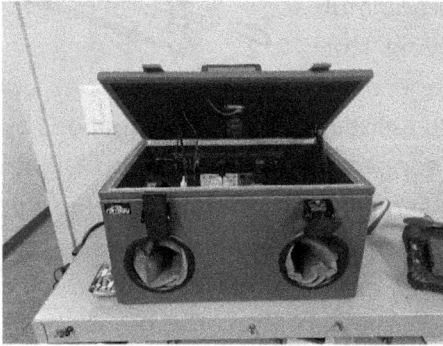

Faraday Box (courtesy of C. McNeil).

Similar procedures are used when examining computers such as laptops, desktops and tablet computers. The computer's data is extracted using forensic methods that ensure the integrity of the original evidence. This can be from a live acquisition of the data or creating a forensic image of all the contents of the device's hard drive or storage area. The data is then examined by forensic software tools to recover deleted data and organize the contents of the drive. This produces a report where investigators can search for specific items that are related to their case.

FUN FACT:

Evidence submitted to the DMF unit for examination may be in an altered or damaged form. This type of casework can involve specialized skills in micro soldering, chip removal, rebuilding damaged devices, or figuring out what a new device actually does. For example, a mobile phone may be purposefully damaged with the intent to destroy evidence or damaged during the commission of a crime.

Both are situations where the DMFE attempts to restore the device to working condition in order to extract digital data. Every device is unique in its complexity regarding accessibility and has different levels of encryption. For example, a device with an unknown passcode typically requires software to try to "break" the pass-

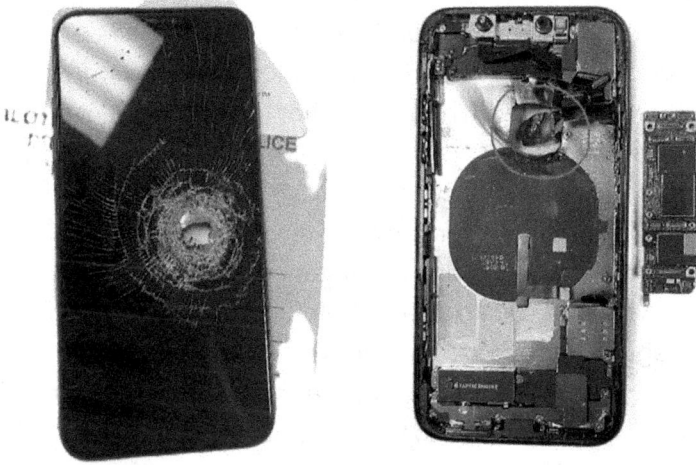

Devices with gunshot damage. (courtesy of C. McNeil)

code. Brute force processing can take minutes to years depending on the complexity of the passcode/encryption.

DMF services are utilized in investigating crimes associated with hacking, network intrusions, and cases involving social media. The DMFE traces IP addresses using law enforcement resources as well as open sources to attempt to determine the potential perpetrators of the crime. Keep in mind, any casework involving internet data requires an examiner to understand and follow jurisdictional regulations. A criminal investigation that is initiated in one state may lead to potential out of state or out of country perpetrators, at which point the investigation may be forwarded to that jurisdiction for follow up.

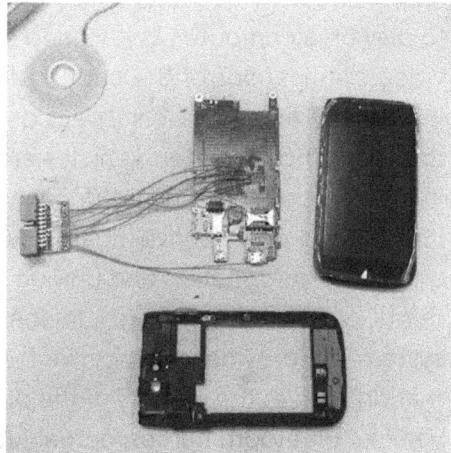

Mobile device restoration. (courtesy of C. McNeil)

Case Scenario 1: Homicide investigation – The DMF unit assists detectives with digital evidence. Often the only link between a victim and an offender is cell phone communication - who the individual last spoke with or messaged. This information can provide investigative leads (i.e., the victim is involved in selling narcotics or guns) and insight into their personal life and acquaintances. An individual's web history and photo/video gallery can provide clues to addresses searched online and reveal images of acquaintances and friends. Metadata associated with images can also be helpful to the investigation. Geo-tagged images or videos are marked with location information which in turn can provide details about where the image was taken, where the individual resides or likes to frequent. Stored location information refers to geographical data stored on the device itself. Some cellular phone manufacturers collect and store location data on where the phone has traveled for a certain period of time. This information can be interpreted by forensic software, oftentimes producing a mapped timeline of the phone's location. During a homicide investigation, video data from Ring doorbell cameras, home surveillance systems, and commercial digital video recorder systems can lead to invaluable investigative evidence. Often these systems are proprietary in nature and the video data collected must be analyzed and converted to play on a computer. With video evidence the DMFE can perform video clarifications of the captured footage. This involves capturing individual frames of the suspect or witness to provide to the media or other officers. When vehicles are a potential source of evidence in a homicide case, the DMFE can view video footage for vehicles of interest and provide investigators with vehicle information like make and model information and unique vehicle characteristics (i.e., color, dents, cracks, damage, etc.). In some cases a vehicle's infotainment system can provide investigative leads. The DMFE can access information such as the identity of the phones that were plugged into the vehicle, Bluetooth names, actual location and map locations of the vehicle, and events of car door openings with length of time since opened. These details can help provide a complete picture of what events occurred in a particular vehicle during a specific timeframe.

Case Scenario 2: Child sex abuse material (CSAM) cases – CSAM cases include investigating individuals that are trading, possessing, and sharing child sex abuse material. These are labor intensive investigations requiring an examination of evidence from multiple pieces of seized evidence such as phones, computers, video cameras, digital cameras, external storage devices, and flash memory drives containing tens of thousands of images and videos. Software is used to categorize and search images for known CSAM material and determine image counts for prosecution. The DMFE can access resources provided by the *National Center for Missing and Exploited Children* to determine the identity of known victims on the seized photographic images. Examiners view the web history and search terms used by the offender as well as any other personal identifying documents' information to determine the identity of the individual using a particular device and what specific content was being viewed online. Establishing the source of the illegal material is accomplished by viewing websites visited, chat groups, and installed applications/programs. CSAM investigations involve a significant amount of legal paperwork such as the application for search warrants and court orders. At times, there can be several processes occurring at both the state and federal level for one case.

Case Scenario 3: Sexual Assault/Rape Investigations – Frequently mobile devices and computers play a valuable role in these complex investigations. Most individuals now utilize apps and text messaging for the majority of their conversations. These can be crucial to establishing the facts surrounding an alleged assault. During an investigation the DMFE may observe activity such as selectively deleting incriminating messages or "baiting." This refers to an individual drawing others in with explicit text messages. Forensic software can recover deleted texts, providing a bigger picture of what actually occurred. Device web history information can reveal search terms used by the phone owner (i.e., how to set someone up, or how much does (insert victim's profession make)). Device location information assists the examiner in establishing a timeline that can in turn be helpful in tracing the victim's steps and corroborating testimony. Images of the victim and suspect can provide evidence relative to their location, physical evidence of contact (for instance, images of the two together or in same location), as well as confirming the time-line of where they were when the image was taken. In some cases the images themselves may be particularly incriminating, for exam-ple, a perpetrator filming the sexual assault or a victim using their phone to collect images immediately after an assault.

Case Scenario 4: Fraud Investigation – Individuals use the internet to commit bank fraud, open accounts, transfer monies, etc. A per-petrator may create fake forms of identification. This activity is often uncovered by the DMFE by detecting templates and images associ-ated with the fake ID's that were stored on the device. Offenders of fraud may take notes or capture images on the device associated with the criminal activity. For example, an employee with access to credit cards who is involved in credit card fraud will simply take a picture of the front and back of a card and save credit card informa-tion in the notes feature on their phone. Data on the use of the stolen credit card (i.e., placing orders online using their devices, shipment of items to their house or work, etc.) can be used to link back to the offender. In addition, these individuals frequently use phone apps to sell or share stolen credit card information to others online and in internet forums. The DMFE can review IP addresses and logon information to detect the location the criminal accessed these accounts. As are the circumstances with other cases involving data access, search warrants are often necessary.

> **BTW**
> Criminals may use technology to exploit individuals online and take advantage of the Internet's anonymity; however, no one is completely anonymous when online. Everyone leaves footprints inadvertently.

Case Scenario 5: Missing persons – These cases often rely on phones and computers for investigative leads. With the proper court paperwork, the DMFE will attempt to locate the individual's phone by tracking the location. Alternatively, devices left behind by the missing individual can be reviewed to scan for contacts, texts, or letters indicating where they may have gone. Information in these communications can provide clues as to whether criminal activity may have been involved or signal an individual's intentions (i.e., move, sever contact, escape, etc.). Sometimes these are the only leads provided in this type of investigation.

The DMF unit also supports patrol divisions. This can include examining video from a convenience store for a petty larceny case to Ring doorbell footage for people passing by. Cases involving illegal drug possession/sale, robberies, auto theft, hacking and other social media related violations all involve using a digital device to commit criminal acts. Similarly, the investigating agency submits any necessary court paperwork for authorization to access and examine these devices and associated data.

Results of a DMF examination involving retrieval and review of video and audio evidence can lead to information that assists other crime laboratory units in their testing. For example, video footage may show a suspect handling certain items which can then be swabbed for potential DNA testing. Audio recordings that are muffled can be enhanced to be understood and may provide information to where events occurred and lead to additional physical evidence or witnesses to a crime.

The DMFE must be capable of adapting to the tools available to the modern-day criminal by keeping up with current technology and trends (i.e., AI generated images and video and digital manipulation of video and documents). As technology changes so do the laws associated with these new technologies. The DMFE must be familiar with the changing legal requirements associated with the examination of digital evidence as the legal decisions evolve to form new or amend existing legislation. This field, like any other forensic field, will continue to expand in the future.

Trace Evidence Analysis

The trace evidence analysis unit (TEAU) specializes in examining evidence that is often
deposited through contact. Contact can
occur by interacting with the scene or individual in some way to cause transfer or deposit
of evidence. The TEAU performs comparative
examinations of evidence samples to known
standards to determine physical matches and
tests liquids or debris used for their explosive or flammable nature.

Forensic scientists in the TEAU support investigations of all
types including homicides, burglaries, vehicular manslaughter,
arson, and sexual assault. Physical comparison of evidence from a
scene to evidence found on a victim or suspect can assist investigators in placing an individual at a specific location or having been in
contact with a specific individual. Trace evidence can also be carried
by the perpetrator from one scene and deposited at a second scene.

Case Scenario 1: Homicide – Shoe
and tire track imprints and impressions are submitted to the laboratory with a request to compare
these impressions against databases
containing reference impressions to
assist investigators in determining
what type of shoe the suspect might
have been wearing and the type of
tire on the vehicle that was at the
scene. A suspect is later developed
and known exemplars from his shoes
and from his vehicle tires are compared to the scene evidence. If suffi-

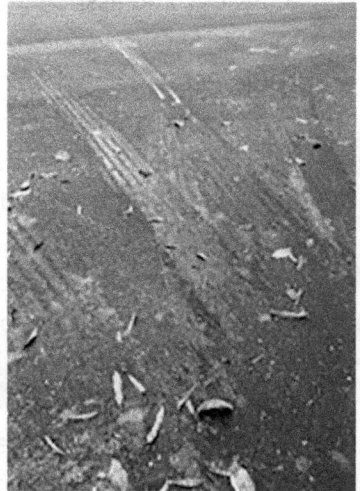

Tire prints from a scene.

cient unique or individual characteristics are present on both the
known and questioned evidence it is possible to determine that a
specific shoe, or tire, left a questioned impression at a crime scene.

Case Scenario 2: Arson – Debris from a fire is sent to the lab to test for volatile compounds to assist investigators in determining possible sources of ignitable liquid residue. The Fire Debris Analysis report lists a class of ignitable liquid and provides some examples of types of products that fall into that class. The agency receives the report and executes a search warrant on the suspect's residence. Liquid products collected during the search, and the clothing that the suspect may have been wearing at the time of the incident, are submitted to the lab for testing for ignitable liquids and ignitable liquid residues.

Shoe print from a scene.

Examples of routine sample types submitted to the TEAU are hair, glass, fibers, paint, shoe prints, tire tracks, tape, building materials, fire debris, and potentially stained clothing, soil, cans, etc., with suspected ignitable liquids or residue.

Because trace evidence can be anything small enough to be transferred without being noticed, microscopes are the main instruments used by forensic scientists in this section. For example, a suspected shooter's hands can be swabbed, and the swabs can be examined using a Scanning Electron Microscope (SEM) for the presence of gunshot residue (GSR) or primer residue. Take note, this examination is different from the type of exam performed in the Firearms unit where the goal of the GSR exam is to determine distance estimations of shooter from target.

Hair and other fibers, paint chips, glass and explosives are some of the materials that can be examined directly with a microscope. In order to get additional information about a sample, the section also uses a variety of instruments, such as The Gas Chromatograph-Mass Spectrometer for Fire Debris and Ignitable liquids. The section also studies the interaction of certain samples such as fibers and paint with different wavelengths of light. An Ultraviolet/Visible Light/Near-Infrared (UV/VIS/IR) spectrophotometer can be used to differentiate fibers such as carpet fibers from two different vehicles. Alternately, the spectra of the questioned fiber can be compared to spectra from a database of carpet fiber reference standards. An infrared spectrophotometer can be used to provide a chemical fingerprint of fibers, paint, or suspected explosive residue. Again, because the evidence items are typically small, these spectrophotometers are usually attached to microscopes.

Trace evidence analysis can be requested along with other crime lab unit services depending upon the type of crime that is suspected, and the type of evidence secured from the scene or individual. For example, in a vehicular manslaughter case airbags from both the driver and the passenger seats could be submitted to the forensic biology unit for blood testing to assist in identifying the driver of the vehicle. The trace evidence analyst might be asked to examine the airbag for the presence of fibers to compare against the suspected driver's clothing fibers. If driving under the influence is also a suspected crime, the forensic toxicologist might become involved in the case to test the blood of the suspected driver for the presence of alcohol. The laboratory sections work together to determine the most appropriate order of analysis, e.g., Case Scenario 2 (above): latent prints, then fire debris analysis followed by DNA.

Firearms Analysis

Firearms analysis is a specialized discipline within forensics that involves the examination, identification, and analysis of firearms, ammunition, and related evidence to assist law

enforcement agencies in criminal investigations. The firearms unit can determine if bullets, cartridge casings, or other ammunition components were fired in a particular firearm.

Firearms analysts, also known as firearms examiners, perform a range of duties to provide valuable information in criminal cases. Due to manufacturing use and abuse, firearms have unique microscopic surfaces. During the firing process the unique microscopic

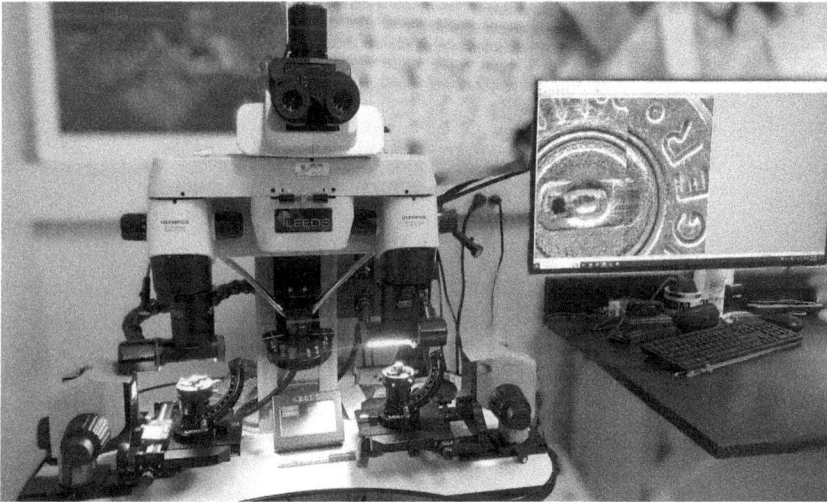

Firing pin aperture shear marks from two fired cartridge cases viewed with a comparison microscope. (courtesy of M. Kurimsky)

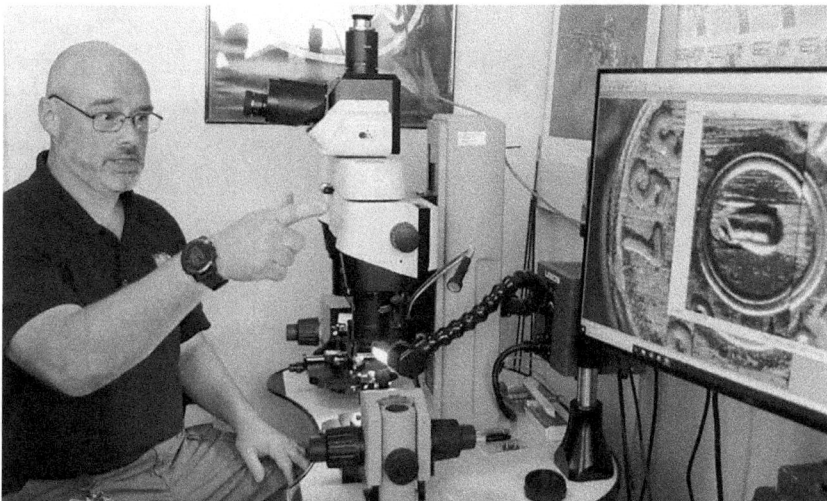

Enhanced imaging of fired cartridge cases viewed with a comparison microscope. (courtesy of M. Kurimsky)

marks present on the firearm are transferred to the ammunition components (bullets and cartridge cases). Firearms identification involves comparisons between fired ammunition components. These evaluations are performed using comparison microscopes.

Comparison microscopes aid the examiner by magnifying two items (bullets or cartridge cases) simultaneously, in separate fields of view, to allow the firearms examiner to evaluate the microscopic marks on each item in a side-by-side manner.

Firearms examiners perform shooting reconstructions. An example of this would be a muzzle-to-target-distance determination test. A distance determination test can involve gunshot residue pattern analysis where both visual examinations of the pattern and chemical tests may be performed to assist in determining distance of the muzzle of the firearm to the target. In the case of shotguns, distance determination can be performed by analyzing the shot or pellet pattern.

Modified long gun with sawed off barrel.
(courtesy of M. Kurimsky)

Firearms recovered from a scene, or seized as part of legislative mandates, can be submitted to the firearm examiner for operability testing. Operability testing tests whether or not the firearm is working (functional). Examiners may also look to see if the firearms are malfunctioning (not working the way the manufacturer intended). Firearms can be modified into an illegal configuration such as shortening a firearm or turning a semi-automatic into a fully automatic firearm.

In other cases, an individual may deface the serial number of a firearm in an attempt to prevent investigators from determining the origin of a firearm. The firearms examiner can perform a serial number restoration making use of physical, magnetic, and chemical processes to attempt to reveal the serial numbers.

Serial number restoration – before and after magnetic processing.
(courtesy of M. Kurimsky)

Serial number restoration – before and after chemical processing.
(courtesy of M. Kurimsky)

The most routine sample types submitted to the crime lab for firearms testing are firearms (handguns, rifles, and shotguns), ammunition, and fired ammunition components (fired bullets and cartridge cases).

Firearms examination provides investigative information utilized in homicides, shootings, armed robberies, and any case where a firearm may be believed to be used in the commission of a crime. In contrast, a firearms examiner may be requested to examine firearms in cases of accidental shootings or self-defense.

The following are examples of typical firearms casework:

Case Scenario 1: Criminal Possession of a Weapon – Officers arrest an individual for selling drugs and recover a handgun from the trunk of the individual's vehicle that appears to have a sawed-off barrel, obliterated serial number, and other alterations. The firearm is submitted for testing to determine its functionality and restore the serial number to evaluate whether the firearm may have been involved in any criminal activity. The firearm examiner can also determine if the firearm is legally owned or stolen.

Case Scenario 2: Homicide – The scene of a double homicide is littered with cartridge casings. The cartridge case positions are photographed and subsequently submitted to the crime lab for testing. Investigators have requested an examination of the cartridge cases to determine the make, model, and caliber of the firearm that may have been used in the commission of this crime. Investigators are trying to link the weapon to a crime scene or between different shootings.

Case Scenario 3: Self-defense – A homeowner claims to have fired his pistol at a masked man who was attempting to force his way into the home. The homeowner claims the shooting occurred at the front door, with the distance between the muzzle of the pistol and the man being less than one foot. Investigators have requested firearms analysis to examine muzzle-to-target-distance to see if the physical evidence is consistent with the homeowner's statement.

Depending on the case scenario, firearms are examined by multiple crime laboratory unit scientists to inspect the firearm(s) for potential evidence of criminal activity (e.g., blood, gunshot residue, fingerprints, etc.). For example, a handgun recovered from a homicide scene will be swabbed for skin cells to attempt to develop a DNA profile to determine who held or shot the firearm. If blood is found on the handgun it will be swabbed and tested to develop a DNA profile to determine the source of the blood. Blood on the muzzle of a weapon could aid investigators in linking the weapon to the scene, shooter and/or determining how close the shooter was to the victim.

The firearms examiner and the forensic biology examiner must understand the nature of the case, the evidence, and the testing requests being made in order to determine the priority of their examination in an effort to preserve all of the evidence. This may mean that the firearms examiner will make the handgun safe for handling by the forensic biologist but taking care not to disturb any physical evidence on the handgun. The forensic biology examiner would then swab the handgun for potential sources of DNA (e.g., blood, skin cells) before the firearms examiner receives the gun for his/her testing.

Fingerprint Analysis

Fingerprint analysis is one of the most consistent and historically established disciplines of all the various forensic sciences and has been used in the identification of individuals for over 100 years. Understanding the uniqueness of friction ridge skin dates back as early as the seventeenth and eighteenth centuries with friction ridge details being observed under some of the earliest invented microscopes. Based on the uniqueness and permanence of friction ridge details, the Latent Print Examiner (LPE) assists in criminal investigations by attempting to identify or exclude a person from having touched an object. The LPE's role is to assist in identifying the source of the latent print. It is the investigator's job to take that information and determine the significance of the identification to a crime scene.

In a Latent Print Unit, the LPE performs the evaluation and examination of friction ridge impressions collected from a surface at a crime scene after a person's fingers, palms, or even feet have made contact. The LPE uses discriminating characteristics such as pattern types, ridge flow, and other minute details called minutiae to compare latent prints to known fingerprint standards to attempt to identify or exclude a person from having touched an object.

Latent Fingerprint vs a Known fingerprint – minutiae points are plotted. (courtesy of A. Lester)

Latent prints themselves are not visible to the naked eye until they are enhanced through physical or chemical processes and can be developed in the field by trained law enforcement or crime scene investigators, or within the confines of a laboratory. Everyone has seen a television show or movie where the LPE is *dusting* for prints. This method of development makes use of powder that sticks to the oils in a fingerprint followed by the collection or *lifting* the print from the surface to preserve it using clear tape or Mikrosil Casting Putty. Other methods of development involve chemicals to either bond or react with components present in the oils of the latent print to transform the friction ridge detail into a hard plastic or apply color to illuminate it. With any of the methods one must take into careful consideration the development and collection of latent prints as they can easily be damaged or destroyed even by the means to retain them.

Latent fingerprints can be associated with many case types. Most common requests for latent fingerprint analysis are associated with property crimes or vehicle theft, larceny and of course homicide. Criminal possession of a weapon is also a case type in which latent print analysis is often requested.

A practical tool used in both the field and the laboratory to document latent prints is the use of a camera. A camera is used to capture the location and placement of latent prints at the scene or on items of evidence. Photographs document orientation of the latent print, what surface it is being collected from, as well as other significant information that can be useful to the LPE when they are evaluating the fingerprint for comparison purposes. For instance, if a person has gripped a coffee cup or has pushed upward on a window to open it. One asks the question, "How is that person's hand or fingers operating the object or surface area?" and

Example of a common surface type and Latent Print Powder. (courtesy E. smith)

therefore, "How are their friction ridge impressions being impressed onto the item?"

In the LPE's world, while the *print* is the evidence, it is the surface type the latent print sits on that dictates the type of processing that is performed. Some of the more common types of surfaces prints are recovered from are glass, plastic, metal, or glossy wood. Latent prints can also be recovered from porous surfaces such as paper and cardboard. The LPE may process items that are wax coated, contaminated with fluids, or present on adhesive surfaces such as tape.

When a latent print is being analyzed, the examiner applies what is known as the ACE-V methodology – this is the Analysis, Comparison, Evaluation, and Verification of the latent print. Determining the suitability of the latent print is the first step, this assists with what further action can be taken with that print. Latent prints with a certain amount of suitability or quality of detail, can be searched against known fingerprint databases at the state and federal level. Every state has their own known fingerprint database or Automated Fingerprint Identification System (AFIS) and the Federal Bureau of Investigations (FBI) has the Next Generation Identification (NGI) database. These databases contain known fingerprint cards (or Tenprint cards) collected from individuals associated with criminal transactions (i.e., arrests) or civil transactions (i.e., applying for a job).

The databases contain millions of civil and criminal fingerprints and are an invaluable tool in the identification of the contributor of a latent print. If an identification is not made upon initial entry and search of the databases, the latent remains in the databases in a specific location or queue referred to as "Unsolved Latent Database" until such time when a known fingerprint card is collected and searched against the Unsolved Latent Database. This has been instrumental in missing person's identifications and unsolved homicides, decades later.

Image of database tool. (Galeanu Mihai, February 01, 2024, iStock-1847709747)

Here are a couple of examples to illustrate where the LPE's skill set is needed:

Case Scenario 1: Vehicle Theft – Several hood ornaments were stolen from cars at a high-end automobile dealership. No tools, gloves, or any other items belonging to the suspect were recovered from the scene. The LPE will examine the hoods of the vehicles to determine if prints were left behind.

Case Scenario 2: Robbery – Two suspects entered an occupied residence and bound the residents with duct tape before proceeding to steal jewelry and cash from the home. The LPE will evaluate the duct tape, jewelry box, safe, and point of entry for potential prints from the suspects.

In cases such as these, often an investigator can attempt to utilize the services of the LPE and the forensic biology unit to help identify the perpetrator of a crime. In this situation the same item of evidence might provide information to both the LPE (impression) and the DNA analyst (DNA profile from skin cells). The LPE must be mindful of downstream analysis, such as DNA testing, in order to determine the most appropriate latent print processing method so as not to affect subsequent testing capabilities.

On the flip side, coordinating examination of an item of evidence can occur in the crime laboratory when a member of the forensic biology unit is performing a preliminary examination and sees what appears to be a latent print. The LPE could be called to the examination room to photograph and process or collect a print.

LPEs go through extensive training to understand the embryological development of friction ridges and the further development of minutiae, including fingerprint patterns and classifications, history of fingerprints, as well as how known fingerprints and palm prints are collected for comparison purposes. By understanding friction ridge detail, the LPE can further use and understand unique biological features and their measurements to identify individuals – this is known as Biometrics.

FUN FACT:

Though iris scans, retinal scans, facial recognition, and voiceprint identification are just some examples of the use of biometrics for identification purposes, fingerprints are still the most common biometric used for identifications in criminal investigations worldwide.

Latent prints and the databases used for identification purposes are used not only to identify unknown fingerprints from crime scenes, but also to identify individuals from mass casualty events. Often, if the genetic material for DNA testing is compromised in some way, the fraction of a fingerprint, or footprint, or palm print, can assist with the identification of a body.

> **FUN FACT:**
> Even when the genetic profile of identical twins is indistinguishable, their fingerprints are unique from one another based on the embryological development, making fingerprints the ideal means of identification.

Questioned Document Analysis

Questioned documents examination, often referred to as forensic document examination, is a specialized field within forensic science that deals with the analysis and examination of documents in order to determine their authenticity and origin or to uncover alterations and fraud. This field plays a crucial role in legal investigations, fraud prevention, and the resolution of disputes involving documents.

The questioned document examiners are called upon to analyze evidence items such as legal documents, bank cards, receipts, test results, hospital records, greeting cards, notes, and letters. For paper documents where mechanical impressions (e.g., typewriter, check writer, rubber stamp) are to be evaluated, the evidence items may often involve the instrument suspected to be utilized in creation of the document in question.

The questioned document (QD) unit will be equipped with microscopes, digital imaging instrumentation, infrared and ultraviolet light sources, video analysis tools and specialized equipment such as electrostatic detection devices (used to detect indented impressions). The examiner may perform a physical exam of an item using visual or microscopic analysis for comparison to known samples, or a chemical exam where ink analysis is needed.

A QD can be an indication or source of criminal activity in many types of crimes including embezzlement, kidnapping, homicide,

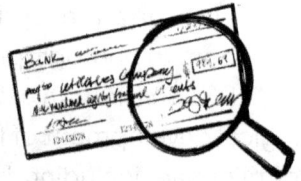

and financial crimes. Forgery could involve evidence such as handwritten documents, photocopies, printing, ink, and paper sources and types. Cases of forgery where comparison of handwriting and signatures require review of a sufficient number of known handwriting samples from an individual because natural variation from sample to sample of the same individuals' writing will occur and no two writers display identical features.

These *samples* are obtained through the normal course of business-executed documents such as old letters, old checks, contracts, etc., or obtained by request where a suspect is asked to produce writing that is similar to that in question. It may not be an exact duplicate, as the suspect may try to disguise their handwriting during the controlled collection of the exemplars. Examiners may request handwriting samples, like the "National Park Tour Letter"[1], where every letter and number of the passage are represented in different combinations to allow the more natural flow of handwriting to occur. These can be modified to incorporate specific words or letter combinations that match the text of the questioned document to increase the chance of obtaining a more true and accurate representation of a writer's handwriting.

PLEASE WRITE THE FOLLOWING PARAGRAPH:

"A tour through our national parks would be enjoyable to you, I know. We left Los Angeles at 7:45a.m., September 20, via Valley Boulevard, and motored to the Grand Canyon in Arizona. From there we drove to Zion National Park in Utah, next a jump to Yellowstone. Then we drove to the coast, into California, and through the Redwood Forest to San Francisco, the commercial Hub, arriving at 9:30 p.m., October 21. Here Mr. and Mrs. John X. Dix of 685 East Queen Street, Topeka, Kansas, joined us. I found the roads good, some quite equal to the best."

A tour through our national parks would be enjoyable to you,
I know. We left Los Angeles at 7:45 a.m., September 20,
via Valley Boulevard, and motored to the Grand Canyon in
Arizona.

Written by: _____   Signature: _____

Witnessed By: _____   Date: _____

Handwriting tool: The National Park Tour Letter[1]

Here are a few examples of case scenarios to illustrate where the QD examiner's skills are utilized:

Case Scenario 1: Check fraud – An elderly man passed away. The family members of the man observed unusual financial activity and alerted the bank. Checks received by the bank for cash were sent to the QD unit for analysis.

Case Scenario 2: Kidnapping – A ransom note was mailed to the parents of a 5-year-old girl that had been abducted while the family was on vacation. The suspect was believed to be a member of a known gang. A typewriter was used to create the ransom note. A typewriter from the suspected criminal was recovered from her employer at her place of business. The QD expert was asked to analyze the composition of inks, dyes, and paper as well as impressions to determine their origin.

FUN FACT

In typewriter cases - sometimes the ribbon (containing the ink used to make the individual keyed letters) will still contain the impression of the words in the message. The ink missing from the ribbon is a result of the letter key's face striking the ribbon and imprinting the ink onto the paper the message was typed on. As a result, examiners have had success "straight reading" the ribbon to compare to and match the threatening letter sent. There are other types of typewriters or electronic word processors that will conserve ribbon and overtype, or not use the ribbon in such a linear manner, making deciphering the text on the ribbon harder if not impossible.

The kidnapping scenario (# 2) is also an example of a case where multiple crime laboratory units may become involved in examining a piece of evidence from the same case. Investigators may request a latent fingerprint exam of the note for potential fingerprints of the suspect. The envelope, particularly the back of the stamp and the inside flap of the seal, could be swabbed for DNA from the saliva of the suspect. All three crime lab units must coordinate their examinations in the appropriate order, utilizing the least destructive method of exam in order to allow for the best testing outcomes.

Multi-unit coordinated examinations are important. The QD examination can be impacted if considerations are not taken prior to assignment of testing order. For example, if a DNA analyst swabs the letter, it has the potential to alter the paper fibers that are imperative in deciphering the indentations, if present. Water causes the fibers to expand, losing the indentations, making it impossible for the QD examiner to visualize later.

If the document goes to a latent print analyst prior to QD examination, the chemicals and powders that are used could make visualizing indentations or writing on the evidence more difficult. A QD examiner, like all forensic scientists, is well-trained in handling evidence appropriately, taking care not to hinder further examinations by other disciplines. QD examiners can take precautions such as using pens to mark evidence that won't interfere with latent examiners performing their work. For example, some inks can interfere with visualization of other evidence elements during an alternative light source (crime scope) exam.

For the QD examination, it is best to have the QD examiner process and photograph the evidence prior to any other examiner handling. The details of the case, the totality of the evidence in the case, and the types of testing available dictate how the agency and the laboratory prioritize evidence testing. In certain situations the agency may forgo the document examination and pursue DNA or latent print analysis if these tests will provide the most valuable information to assist the investigation. Where possible, it is best to account for all potential testing, making decisions based on an attempt to retrieve any and all probative information.

Want even more information? Visit a crime laboratory web site.

Some crime laboratories post informational videos where working scientists talk about their jobs. Search the agency websites to see firsthand what these careers are all about.

Get Personal

Starting point: define personal interests and values

If memory serves me correctly, somewhere around the midpoint of my junior year in high school the guidance counselor held one-on-one sessions to discuss student's interests. I say *the* guidance counselor because we only had one.

First, we took a test to survey our interests on a bubble answer sheet paper.

Vintage Scantron Image [2]

SIDE NOTE:
Sad fact for this child-of-the-1980s, I learned that most schools are phasing out the Scantron - the machine responsible for tabulating bubble answer sheet results.

Next, we met with the guidance counselor for a follow-up session to go over the results of our interests test. But can I be honest – no offense to my guidance counselor – those tests were not at all helpful in narrowing down a career path.

No singular test could ever capture the *essence of me* and my *future destiny.*

A bit impersonal. Right?

To my recollection, my test results revealed an elevated *interest* in nature, so naturally this is how this went down...

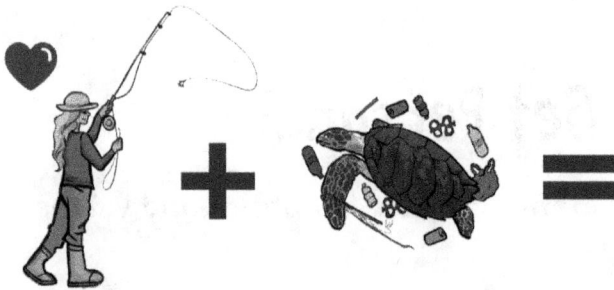

Fishing + Environment = Marine Biologist

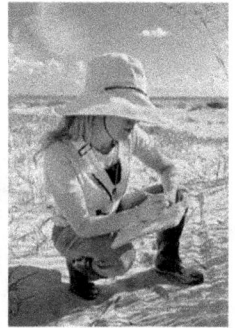

Well, obviously!

Now do not get me wrong, I loved my undergraduate education in marine biology. But it was not until a couple of years after graduating, steeped in homesickness for my first loves (camping, fishing, hiking, all things NATURE), it became clear that one of my favorite parts of my education had more to do with the desire to piece information together as was evidenced during moments of exploration. Combing the beach for example, stumbling upon a dead and bloated carcass (fish, crab, bird), pondering the cause-of-death. That is when things got really interesting for me! That's about the same time I came to discover the field called: forensics.

My point: My career path was NOT 'obvious' to me during high school. Partly because I should have dedicated more time to learning about myself and my interests. Partly because I would have benefited from resources to help me do just that.

To hammer home my point, interests only scratch the surface in helping you determine what translates into a potential vocation. You need to tap into your experiences and understand what they have to teach you about your passions (your heart).

"...incline your ear to wisdom and apply your heart to understanding..." (Proverbs 2:2 NKJV)

TIP: If you want to be a scientist begin to think and act like one. Collect as much data as possible to inform your decision and thoroughly test your hypothesis.

Teasing out from an *interest* what brings you joy, holds your attention, syncs with your values, and taps into your unique gifting - well that is a bit more personal and a heck of a lot more useful in steering you toward a fulfilling gig.

In this chapter we will begin by defining your interests through your values and then we will loop back to look at what makes you "you," your personality. Later on in the book we will explore your strengths and begin to connect the dots by investigating how each puzzle piece (interests, values, personality, strengths) fits together.

SIDE NOTE:
If you enjoy puzzling, you might find that you rank highly in the critical thinking skill set where discernment and solution-based thinking are strengths of yours (also traits commonly attributable to scientists)

Interests and Values

What brings you joy? What holds your attention? What matters to you? These are a few of the questions we want to answer in order to understand what we favor and what we value.

How do we define what we value? Our basic convictions, whether generational or cultural, inform our decisions and motivate us to engage in meaningful work. We find meaning in our work when we uncover purpose.

We can begin to get at this information by performing a bit of a self-evaluation. For each of the tables below (Steps 1-4) try to list your top 5. It may be helpful to list out all that comes to mind for each table first on a separate paper and then rank them to get to your top 5.

Remember – this is *not* about listing activities or qualities that might meet the approval of society or fit in with current day culture's opinion of what is worthy, this is about you and what you are enthusiastic about. Focus on you!

What captures your interest? Begin to think... what could I talk about or learn about for hours each day and not get bored?

If you are having trouble with this exercise you may choose to tap into the minds of those who know you best to help you tease out this information. Likewise, *after* you have tried this exercise on your own, seeking input from family and friends can also be a clever way to see how you perceive yourself versus how others might describe you. It is quite possible that you do not give yourself credit for your talents. There may be strengths you possess you did not realize you had.

Let's get started.

ACTIVITY: Self-Evaluation

STEP 1. Spend a few minutes thinking about subjects that interest you. *NOTICE: I have pre-filled each activity with examples to guide you through the process.*

My favorite high school subjects are:	The things I like most about this subject are:
environmental science	learning how pollution affects the environment
biology	studying life systems and causes of human impact
chemistry	performing experiments in the lab

STEP 2. Consider what it is that you do, produce, or create that routinely brings you a sense of satisfaction. What have you done where you have felt the most effective, successful, useful, or valuable? We all have activities and accomplishments that we are proud of. It is ok to brag about yourself.

I do well at:	This brings me joy because:
fixing things	I like figuring out how to solve problems
researching	Learning how things work and how to care for them
participating	I do my part to help out

I'm certain there are specific topics or issues that charge you up...

STEP 3. Consider what is the most meaningful to you by investigating your values. What inspires you (stories or articles that pique your interest and draw you in)? What type of behavior do you respect? Alternatively, consider what makes you angry. Anger can illuminate values born from areas or issues you hold concern for.

I am concerned about:	This is evident in what we say and what we do. List behaviors that demonstrate your concern:	Therefore, I value:
reducing waste	recycling, composting, buying used items	environmental care

And how about your daily activities...

STEP 4. Look at your day-to-day activities – how do you spend your time? SUGGESTION: If you can't think of an activity to list in the '*I spend my free time...*' column, then use this column to write down responses to this question: What could you do all day and never get bored?

I spend my free time:	I am motivated to do this because:
reading gardening books	Growing vegetables at home offers me healthy, chemical free food and provides an opportunity to share with my neighbors.

Great start! Now let's pull it all together.

STEP 5. Check Yourself – In this section you will review and compare the information in the previous tables considering the following questions:

Do the professed values (things you care about) align with your actual behavior (activities you do in your free time)?

Are there any shared or similar values or behaviors associated with the subjects you most enjoy learning about?

Do your brag list activities (areas you indicated you succeed in) match up to topics in your top five academic areas?

TIP: Common themes indicate true interests and reveal where your talents live.

With the goal of identifying *patterns* and *overlapping themes* from the information you listed in the tables in Steps 1 – 4, follow the instructions in each category of the chart below and place your response in the area provided to the right of each category..

SUBJECTS

Review the *'things you liked most'* about your favorite subjects. These subjects hold your attention. List any words or phrases you wrote down more than once or are similar to each other in theme as a **reason for capturing your interest**.

ACTIVITIES

Review your answers *for the joy you feel* when you engage in the activities you are proud of doing. List any words or phrases you wrote down more than once or that are similar to each other in theme as a **reason for sparking joy.**

CONCERNS

Review the topics that concern you. These are the causes you want to be involved in, followed by the *action you take to demonstrate your concerns*. List any **behaviors or values** you wrote down more than once or are similar to each other in theme.

TIME

Review your daily pursuits. List any words or phrases you wrote down more than once or that are similar to each other in theme as a **driving force to motivate you** to do these activities over others when you have a choice.

THEMES

Review your answers in each box above (subjects, activities, concerns, and time). List any **common theme** or **shared pattern of thought** you can identify from your previous responses.

Now that you have a bit of self-awareness about your interests, let us investigate how your personality (natural temperament) plays into *work* that excites you.

4

The 'Character' in the Person

What do your personality traits reveal about your values and behaviors?

The Personality Assessment

The self-evaluation exercise we just walked through in the previous chapter is a terrific way to begin to understand what appeals to you and what you care about. When you are engaged in these thoughts and activities you feel fully alive. Energized. Motivated. Hopeful. Why? Because passion is fueling purpose.

Continuing down this road of self-evaluation we can parse out specific traits that reveal *why* it is that you find joy in the activities you listed. Your hard-wiring.

You might be the only one in your family that is curious about creepy, crawly things – you have an ant farm, pet snail, or collection of specimens and now you can envision parlaying that "hobby" into a full-time gig of sorts.

A forensic entomologist studies insects and how they interact with a cadaver to assist law enforcement in a criminal investigation.

Bugs + Dead Things = Forensic Entomologist

Simple. Right?

While it is pretty easy to imagine that a person who runs from the sight of a cockroach might not be the best *fit* for a career as a forensic entomologist. Quite the opposite may be true of the bug lover-collector-investigator type – bugs are cool and even cooler on dead things!

This should build the case for trying to make a bit of additional discovery. Learning how your unique personality traits reveal jobs you might enjoy more than others is a super valuable pursuit. But how?

We will get there, but first – we have to find the underlying cause of *why* we enjoy doing specific activities and *why* we might be more successful at doing some rather than others. This is revealed in aspects of our personalities that are hard-wired, pre-disposed; having to do with how we feel about doing certain things, the type of environments we are comfortable in, and how we behave in different situations.

For example, a person who prefers to use facts, data, and logic to solve problems, may not flourish in a role where they are required to sense other's feelings, such as the role of a therapist. This is where the personality assessment can provide meaning behind the *why* we are more inclined to favor or behave a certain way.

There are a number of personality assessment tools that you can find on the internet. These tests consist of a series of questions to assess your preferences and behaviors for processing information, relating to people, decision making and shaping your life. The assessments we will use for the following activities are loosely based on the Jung typology [3] – a method developed from Carl Gustav

Jung's theory that behavior is not random but the result of how people receive and categorize information and then draw conclusions from it. There are two *attitude types*, (extrovert and introvert)[4] and personality patterns falling under four psychological functions (sensing, intuiting, thinking, and feeling). Modern methods are based on Jung's original theory but are broken out into a larger number of personality types.

Examples of these personality-type assessments are:

- Myers Briggs Type Indicator (MBTI) Test (psychological preferences)
- Jung Typology Index (cognitive functions behind preferences)
- DISC Assessment (behavior style, communication style, work preferences)
- The Enneagram of Personality (core *type* and behaviors associated with it)

Examples of websites where these assessment tests can be found:

- https://www.16personalities.com/
- https://www.truity.com/
- https://www.123test.com/
- https://www.thepersonalitylab.org/

Most assessments take about 10-15 minutes to complete. Many are free and others charge a fee for a *full report* so take a few minutes to find assessments that best suit your needs. For maximum benefit, try to find assessment tools that will provide you with a **personality type** and information about specific **strengths**.

Alternatively, if your assessment results do not include a *strengths* section you may be able to reverse engineer your results to find a general example of the strengths attributed to the personality type the test assigned to you. You can do this by searching the tester's website for its description of the personality *type* assigned to you. **Within the website's explanation of the *type* you should find a general list of strengths specific to that *type*.**

Heads up! The associated *strengths* will be used in a subsequent activity so do not skip over this part.

TIP: You will want to take your assessment tests on different days to give your mind and emotions the best opportunity to provide thoughtful responses. Set aside uninterrupted time where you are energized and most likely to be in a neutral frame of mind to minimize the potential for fatigue, bias, or misinterpretation.

ACTIVITY: Assess Your Personality

STEP 1. Grab a notepad and create 3 large rows (similar to the format in my chart: Syncing Up Your Personality Traits).

STEP 2. Find a personality assessment test and complete it (*assessment #1*).

STEP 3. Review your results.

STEP 4. List the *key* traits from *assessment #1* in the first row.

STEP 5. Complete a *second* assessment test (new assessment tool – different source). List the *key* traits from *assessment #2* in the second row.

STEP 6. Complete a *third* assessment test (new assessment tool – different source). List the *key* traits from *assessment #3* in the third row.

FINAL STEP. Circle traits that are identical, similar, or common in theme among the results you listed in all rows. In my example I used different circles (solid, dashed, dotted line etc.) to group traits of similar themes together. You could use colored markers.

Finding similarities amongst the results of three separate tests should give you more confidence that you have identified your true traits and strengths.

SYNCING UP YOUR PERSONALITY TRAITS

Circle traits that are identical/similar/common in theme

Introvert (fewer social interactions, calmer environments)

Assessment 1

Test Name:
16 Personalities

Thinking (impartial, logical)

Observant (strong focus on what is happening or about to)

Judging (decisive, thorough, organized, clarity, predictability, structured)

Intense ability to concentrate

Serious and direct

Assessment 2

Test Name:
Enneagram

Natural talent for finding solutions

Practical and frugal

High internal standards

Hardworking and diligent

Plans and schedules for efficiency and optimization

Efficient and methodical

Enjoys working alone

Assessment 3

Test Name:
DISC

Focus is on accuracy and precision

Orderly and exacting

Prefers well defined tasks

If you are curious, here are traits commonly associated with science *types:*

attention to detail	methodical
objectivity	work well with others
integrity	analytical skills
honesty	good written & verbal skills
independent	careful
problem-solving skills	resilient
thorough	able to meet deadlines
critical thinking skills	unbiased

Now take a look at the traits in your chart. Do you see any similarities between the traits in your chart and the traits listed above?

Well, if it was not obvious from this quick comparison – each of us is unique. We have just confirmed that scientists, just like everyone else, have not been popped out of a prototype or pressed out of a mold. Rest assured that even if none of the traits above aligned with those that you listed – As individuals we will be unique in our abilities and passions. There is no exact fit.

As you spend a few minutes thinking about the values and behaviors that have been revealed to you in the previous exercise you might begin to categorize them into either *good* or *bad* or *positive* or *negative.* Try not to do this. Instead, choose to find encouragement in identifying how your traits and personality can be purposeful for you.

That said, this is a great time to explore your **strengths** to see how they might play out in the day of a forensic scientist. And hey – you might discover that what you labeled as a *bad* or less desirable trait is actually a gift.

Let me demonstrate in the following chart. On the left are my strengths taken from one of my personality assessments. On the right are my thoughts on how each strength helps me function in my daily work as a forensic scientist.

Strength	How could this serve me in performing **forensic science** casework?
Honest and Direct	Truthfulness and trust are a big deal. This work involves the law.
Disciplined	Testing can take hours to days; it requires persistent attention.
Extremely Responsible	Violent crimes present immediate threat to public safety. Being reliable and committed to my work helps me deliver results on time.
Calm and Practical	Keeping a clear head allows me to stay objective and make rational decisions based on data and facts.
Organized and Effective	Being prepared and working in an orderly manner produces quality work and limit s my likelihood of creating errors.
Research-Oriented	Using valid techniques to gain knowledge from the data provides factual information rather than assumptions.

Now you try...

Put on your lab coat and picture yourself in the crime lab. Your supervisor has handed you a case. You've got to test the evidence and write your laboratory report. Your results need to be sent to the law enforcement agent for review. The results of your work could provide information that might help assist with the investigation and lead to solving a crime.

ACTIVITY: Strength Training

STEP 1. On the first line, enter the type/name of the personality assessment tool used to define your strengths (e.g., 123test, truity, etc.).
STEP 2. List your strengths in the first column.
STEP 3. Consider how each strength might serve you in the field of forensic science. Write your response in the second column, next to the associated strength.

My strengths (Assessment tool name: _____)	
Strength	How could this server me in performing *forensic science* casework

Congratulations! You just used science-y skills... you *reviewed* the *data* and allowed it to *inform* your *conclusions*.

Your strengths should be a source of confidence for you. Later on, in the **Prepare to Launch** chapter, you will have the opportunity to incorporate your strengths into your résumé. Get excited!

Let's move on to investigate your attitude toward *work* in relation to career fitness.

Identify Your Fit

Aligning your interests, values, and strengths to career paths

In the previous chapter we discovered our personality types and traits unique to our interests and values. If you are like most people, upon first becoming aware of our traits we tend to focus on the negative aspects. However, it is my hope that having done the previous exercises you gained a new perspective. That is that you are accepting of both the praiseworthy and the perceived negative traits you uncovered about yourself. Armed with the knowledge that the more you understand, the more agency you have to choose how best to utilize these traits to your benefit (passion AND purpose) and to the benefit of others (vocation).

Areas where you may find things come more naturally to you – a strength – and more importantly, areas where you tend to struggle (AKA opportunity for growth).

TIP: Often it is the struggle that provides the training that produces the trade.

So do not allow your struggles to minimize but rather to redefine your opportunities.

What are your opportunities?

The last step is to connect the dots. Your personality not only shapes your interests but also your attitudes. Attitudes shape behaviors. In this chapter we will explore behaviors – your behavior toward others, your environment, your work tasks – and how these behaviors impact your ability to flourish.

Flourishing – where we find engagement, meaning, and application of our abilities.

Many academic institutions offer a career assessment tool on their website. It is designed to help prospective students find the educational program that is right for them.

The idea here is to perform a fit assessment. The goal - to discover how key factors of your personality predict your preference for and fulfillment in a career area. We will use the Holland Code System for this final evaluation.

While you may be thinking... *What the heck? Another test - I am all tested, assessed, and surveyed-out.* Trust me, you want to understand more fully how best to select a vocation where you are *fit* to flourish. I suggest investing time in this one final tool because not only will it inform the conclusions you are forming about your career preference, but it will also come in handy when you build your résumé.

Hey procrastinator - you will need a résumé NOW, during, AND before you finish college! We will chat more about this in Chapters 9 and 10. Hang tight.

The basis of the Holland Code System is to associate your work interests to a list of related work tasks and activities that you might be drawn to, as well as generating a list of careers that align with those interests. The careers line up with one of the six work-personality types: realistic, investigative, artistic, social, enterprising, or conventional. The idea is that an individual working in a career that matches his/her personality-type, and preferred tasks, will be content and successful.

You can find several websites that offer a test based on the Holland Codes. Some are free and some are not. Below is one example of a website offering such a test but there are others available. https://www.truity.com/test/holland-code-career-test

Activity: Career Assessment

STEP 1. Take a few minutes to find and participate in a Holland Code Career Test assessment. Once you have your results, take a look at your career *interest profile* and pay attention to your highest-ranking *interest profile*.

> SIDE NOTE:
> This former forensic scientist scored highest in the area of realistic/builder with a secondary high score in the area of investigator/thinker.

STEP 2. Review the results. As you review your results, ask yourself: Does your profile make sense to you? Does it fit?

For example, if you ranked highest as a...

Realist type, ask yourself: Do I prefer machines, objects, tools over interacting socially? Am I mechanically inclined? Do I like well-ordered activities?

Investigative type, ask yourself: Are you generally curious? Do you seek learning over simple dull activities?

Artistic type, ask yourself: Do you want to be free to create rather than follow a script?

Social type, ask yourself: Do you seek to be helpful, to inform, to serve?

Enterprising type, ask yourself: Do you enjoy achieving? Are you comfortable being in charge or speaking in front of others?

Conventional type, ask yourself: Do you find comfort in following an orderly routine or method?

Be sure to check out *each* of the Holland Code *interest areas* to see how closely your results reflect your personality traits relative to each interest area.

Realistic: practical, structured, independent, sensible, mechanical, traditional, down-to-earth; may enjoy jobs with tools, machines, or using a physical skill

Investigative: intellectual, curious, logical, analytical, scholarly, independent; may enjoy jobs with theory, research, and intellectual inquiry

Artistic: original, creative, independent, intuitive, sensitive, imaginative, spontaneous; may enjoy jobs having to do with art, design, language, and self-expression

Social: compassionate, patient, helpful, friendly, generous, cooperative; may enjoy jobs assisting, teaching, and serving others

Enterprising: assertive, energetic, confident, ambitious, adventurous; may enjoy jobs leading, motivating, and influencing others

Conventional: orderly, precise, detail-oriented, conservative, thorough; may enjoy jobs managing data, information, and processes

   Truity Blog [5]

Ok, don't lie, I know you peeked. What did the *Career Match Section* of your assessment reveal about the one-and-only you?

STEP 3. Analyze the *Career Match Section* of your Holland Code Career Assessment Test.

This exercise is intended to help you learn if you would be a good *fit* for a career in an investigative science. It is helpful to know if your assessment revealed career categories that have a foundation in science and math.

Review the list of potential "career matches" with these questions in mind:

Question 1. Do you see any careers in science?

Question 2. Do you see any similar skills, education, or experience across the different careers (e.g., independent, analytical, research)?

If the answers to the above questions are yes, then you should feel more confident that your data is supporting a good *fit*.

If not, do not be discouraged. All of these tools you have been using are simply providing data for you to evaluate to help you find the best *fit* for you.

Hey, if there are other careers that popped up in your report that spark your interest – by all means – this is an opportunity to look into those too. An education is a huge investment. Explore as much as possible now to inform your decisions about your future.

If you are not convinced of a good *fit* or are overwhelmed by too many options that seem to *fit*, you may wish to consider taking all of this data and knowledge to a professional. A career counselor can probe even deeper to help you sort through your data and give your more clarity about *fit*, abilities, and feasibility.

Remember: No test can account for placing you into the exact one and only career because it can't account for all the facets that make up the multi-faceted 'you.' This is why you are doing the 'work' in this book to gather as much information as you can to make an *informed* decision about your career choice.

A Note about Crime Lab Organization and Workflow

Depending on the crime lab's staffing model, a laboratory technician (not a forensic scientist) may perform a subset of tasks under the direction of a supervisor or forensic scientist. For example, the technician could prepare solutions, stock reagents, and perform all quality control testing while the forensic scientist performs the testing of the evidence. Alternatively, laboratories could task the forensic scientist with the initial evidence exam and preparation of the evidence for testing while tasking the laboratory technician to run a subset of the tests. In this model, the forensic scientist steps in to analyze the data from the testing and writes the laboratory report of analysis.

Other laboratories might have multiple individuals involved in the process depending on their level of training or technique-specific training. In some units of the crime lab, a Forensic Scientist may be trained on one technique at a time and therefore can only perform certain steps in the overall examination and testing process.

Larger laboratories might structure their testing around rotations – this is where a scientist trained in several or all techniques could perform one technique for say a month (on his/her cases and other scientist's cases in the batch) and then perform a different part of the process the next month, and so on, on a rotating basis. Still, smaller laboratories with fewer resources and casework demands might organize their work around a start-to-finish model where the Forensic Scientist works a case from beginning (examination of evidence) to end (drafting report of findings).

Like the crime lab, the potential for varying workflows and level of individual employee involvement is true of many vocations. Yet there are still other factors to consider when thinking about how a person *experiences* their daily work.

Job Characteristics

So tell me how you really feel?

When you hear the word *work*, does it make you want to crawl under the table and hope that your parents, teacher, or manager do not notice you?

There may be many reasons why you either jump at the chance to take it on, hesitantly raise your hand to volunteer, or hope you have done your best to camouflage yourself into the background where no one can call on you to participate.

Could your response be more reflective of the *request*???

- type of task
- level of involvement
- level of independence

When you *work*, you invest your physical, mental, and emotional energy.

Your desire to do work has everything to do with how you view work. Your attitude surrounding the term **work** has a lot to do with whether you are likely to engage in it. If you view it as a chore or burden, then you are likely to either avoid, procrastinate, or slug your way through it (not giving it your best). But if you view it as an opportunity, privilege, or an offering of service, you are likely to step into it willingly and with a spirit of presence and enthusiasm.

Your likelihood for engaging in any activity has everything to do with how meaningful the work is to you, the demands of the work, the characteristics of the work, access to resources, level of participation, and so on.

Every career has a set of defined characteristics as to how the work will get done and the category of work.

For example, when I'm baking cookies, I don't make all of my ingredients from the raw materials (flour from the wheat kernels), because for me, I don't have endless <u>hours</u> nor all the <u>equipment</u> and <u>resources</u> necessary to create my ingredients. Not to mention how worn out I would be after that <u>time investment</u> that I would not enjoy the baking process.

No – I begin at the gathering ingredients stage. Then consult the recipe and get straight to the measuring and mixing.

There are also aspects of the baking process that I <u>prefer to do alone</u> because these steps <u>require more of my focus</u> (measuring & mixing of ingredients and preparing the dough) versus together (placing on the cookie sheets, decorating/frosting and eating).

Similar to baking, a forensic scientist requires materials and equipment to examine a piece of crime scene evidence. The forensic scientist is responsible for the work tasks assigned and the skillset for which he/she is trained and proficient to perform. Crime laboratories design and optimize their workflow based on the number of testing requests they receive and the available resources (trained personnel, space, equipment, etc.).

> On second thought... this last point is debatable. I do love my sweets! However... I always employ at least one or two quality control testers before serving. And yes – that is an official title!

With task types and workflows in mind, let us consider what it is that you prefer when it comes to doing the actual work. Do you tend to enjoy some tasks versus others? Go the bonus extra mile and do the work below to find out.

Bonus 'Work' (hint, hint – before starting, check your attitude. Are you enthusiastic about this extra work? Are you just planning to slug-it-out? Cheerful outlook now – this is you working *for* you!)

 ## Activity: Job Fit

Take the survey on the opposite page. Circle box A or B in response to each of the five (5) job characteristics..

When you are performing a job where you learn you have performed well on a task you were responsible for, and care about, then you are more likely to be operating from a high level of motivation, performance, and happiness.

If the majority of your selections were in the boxes labeled 'A', job satisfaction (for you) is likely enhanced in roles where you experience meaningfulness, are responsible for outcomes of the work, and are aware of the results of the work activities.

This is true of forensic scientists. However, should you settle on a career in the forensic sciences, you will find a mix of job characteristics from both columns (A & B). You will be challenged. A forensic scientists' work is not the work of a maverick. You do not take risks with evidence, you use prescriptive, validated processes and your work is always checked. This work is exciting! Every case presents a new set of information, an element of the puzzle that you get to help solve. You will feel the impact of your work. It is rewarding! Gratifying!

Convinced you are headed in the right direction? Continue on with me as I share the intel on 'what you need to know' about getting your foot in the door at the crime lab.

JOB CHARACTERISTICS

Circle your preference for each (A or B)

B
Low Variety

A
High Variety

Skill Variety - number of different activities required; varying skills and methods

A
Start to Finish

B
Part of the Whole

Task Indentity - requirement to complete a part of the task or the whole job

A
Highly Significant

Task Significance - level of impact the task has on other people (on other people's work or lives)

B
Least Impactful

A
Select & schedule the task to be performed

B
Follow a prescribed standard procedure

Autonomy - level of freedom and independence offered (in relation to selecting & scheduling the task)

A
Work is completed but rarely requires inspection

Feedback - amount of information supplied about one's performance as a result of carrying out the work activity

B
Work is checked reguarly

**Concepts in the Job 'Fit' Activity are based upon the job characteristics model (JCM) 6.

Reality Check

The internal workings of the crime lab, cross–over into the legal arena, and the quality system

There are some things you should know about getting your 'foot in the door.' In this chapter we will peer inside to look at the crime laboratory operations through the lens of the standards that govern it.

Whoa, back up – Did I lose you? Let me explain. *Standards* simply refer to the set of conditions established to promote and attain a level of quality.

The facilities of a crime laboratory are unique to its location, association, supporting agencies, and capabilities. Laboratories can function in various capacities offering services at a national, state, county, or city setting. Some crime laboratories are supported by a local governing body such as a state or county government and others by a law enforcement agency (i.e., police or sheriff's department). The laboratory leadership may fall under a military, paramilitary, or civilianized workforce structure. Meaning, the lab could be staffed with employees who have law enforcement training or employees with no law enforcement training (often referred to as having a civilian background) or both.

No matter the agency association or structure, crime laboratories as a whole (facilities, testing, personnel) are held to extremely high *standards* due to the fact that the services they provide have a direct impact on the criminal justice system.

Reliability

Forensic scientists in the modern-day crime laboratory are hired to perform scientific analysis to assist in criminal investigations. The scientist and their work must be reliable. The laboratory will ensure that the scientist and their work are trustworthy through several means:

- ✓ Hiring process
- ✓ Training
- ✓ Competency testing (administered after training is completed, qualifying the scientist in a procedure/method)
- ✓ Proficiency testing (annual check to demonstrate the scientist's abilities to accurately and properly apply procedures/methods)
- ✓ Technical review (2nd review by another qualified scientist)
- ✓ Quality assurance & control measures
- ✓ Annual audits
- ✓ Accreditation

Standards

Standards and organizations that develop standards have been in place across many businesses and scientific areas for as long as I can remember and certainly before you were born. The forensic industry is no different.

Today there are standards covering all aspects of a forensic laboratory system such as – facilities, operations, testing, and personnel. Further still, there are standards that apply only to a specific field of forensic science, for example:

Forensic Biology unit's DNA testing - The *Federal Bureau of Investigation* (FBI) produced *Quality Assurance Stan-*

dards under the Scientific Working Group on DNA Analysis Methods (SWGDAM). In order to perform DNA testing on evidence samples the laboratory and its technical personnel responsible for testing must meet these standards.

Latent Fingerprint unit - The *International Association for Identification* (IAI) produced standards in the science of fingerprints and established a *Latent Print Certification Program* to validate a latent print examiner's expertise.

Firearms unit – The *Association of Firearm and Tool Mark Examiners* (AFTE) produced standards and a certification program to demonstrate knowledge and skills among firearm and tool mark examiners.

In the earlier days professionals within the forensic science field came together to discuss and promote standardization of practices. These discussions eventually created an awareness in the field of the need for the development of standards specific to the forensic science laboratories.

One example of a community whose discussion transformed into a standards-generating-body is the story of the *American Society of Crime Laboratory Directors (ASCLD).*

ASCLD formed to create a community where forensic interests, principles, techniques, and practices could be discussed. As technologies advanced there was a deeper need to provide the larger forensic community (laboratories, academic institutions, and practitioners) with a means to assure validity of methods and analysis that would be admissible in court. The result was the formation of the *ASCLD Laboratory Accreditation Board (ASCLD/LAB)* to produce **standards** and a means to establish adherence to these standards in the form of a formal inspection (AKA - an **audit**). The outcome of the *audit* either grants or denies a laboratory or laboratory discipline the ability to perform analyses (this is what is referred to as **accreditation**).

Accreditation is the formal recognition by an independent entity using technical experts to assess and conclude that the laboratory is operating according to the standards it is being measured against. The application of standards and the accreditation process serve to provide the criminal justice community confidence that the forensic services provided by the laboratory are fit for purpose and maintained at a high level of quality. Standards do not replace the crime laboratory's validated procedures, methods or policies as the forensic practitioners continue to determine the appropriate method to apply to a particular process.[7]

Today, laboratories performing forensic testing in the United States generally seek accreditation under the requirements of the *International Organization for Standardization/International Electrotechnical Commission (ISO/IEC) document 17025.*

However, the development of forensic science-specific standards has evolved and expanded in the last decade to cover nearly every discipline and sub-discipline and include standards guiding the crime laboratory's scientist/examiner training programs – enter the *Organization of Scientific Area Committees (OSAC) for Forensic Science.* The OSAC's were established to draft best practices that would effectively reduce the variability seen across the forensic science disciplines in the following areas: training, techniques, methodologies, reliability, limitations, research, general acceptability, and published material.[8] This is an open process where all interested or affected individuals and groups can participate by working with the OSAC's to draft standards via consensus. Once drafted, a standard is subsequently posted for public comment (another opportunity to establish consensus) before being sent to a standards development organization, such as *Academy Standards Board (AAFS ASB)* to advance and implement them.

Why is this important to you? Great question.

As you'll remember, the goal of standardization is to make sure the proper training and framework are in place to produce reliable testing results. You can think of these in simpler terms as two larger categories of framework to support best practices in forensic science:

Scientific standards

In the field of forensic science, scientific standards exist to confirm the reliability of the scientific principles and techniques used to analyze and interpret the evidence. The procedure and methods must be valid and scientifically accepted within the scientific community.

Examples of scientific standards applied to forensic science casework are:

- forensic scientists must be trained, competent and proficient in the discipline/method they are able to perform
- new method must be validated (prove no deleterious effect, appropriate application) prior to use in casework/on evidence
- reagents that are critical to the method and its outcome must be quality control checked prior to use

Ethical standards

In the field of forensic science, ethical standards exist to provide a level of conduct that ensures an accurate representation of the truth, promoting the qualities of integrity, impartiality, and confidentiality.

Examples of common ethical standards for forensic scientists are:

- accurate representation of credentials and qualifications
- accurate representation of findings
- impartiality of examination and testimony

Falsifying records, misrepresenting qualifications, and omitting information that would favor one side versus the other is considered unethical. The need to assure the public's trust in the credibility of the work of a forensic professional and the crime laboratory is paramount.

From the vantage point of standards and integrity, the quality system serves as the backbone of the crime laboratory structure. With this knowledge as a foundation, you are prepared to understand the daily internal workings of the crime lab and duties you might not have otherwise anticipated of the forensic scientist until now.

Understanding the Crime Laboratory Environment

If you've taken a class that involved laboratory work, then it should come as no surprise that the work environment of a forensic scientist is indoors.

Reality check #1 If you imagined you'd be working outdoors, at the crime scene, then, "Houston, we have a problem," and you might want to reevaluate career paths.

Windows

Author's previous workplace – the crime lab in Albany, New York

SIDE NOTE:
Most exterior walls have windows – so you won't necessarily be vitamin D deficient. Get your vitamin 'N' in by going for _Nature_ walks during break times.

But hey, aside from the physical aspects of the facility and the amount of sunlight you may or may not receive, now is a good time to get perspective on other aspects of the work of a forensic scientist. Especially tasks that can seem less glamorous than TV shows like *CSI Miami* have led you to believe.

There is no shortage of work in a crime laboratory and the scientist's day-to-day can be filled with a variety of different tasks depending on the discipline. You'll recognize the following list of job duties from a previous chapter.

- Cleaning laboratory benches and shared areas
- Preparing solutions and stocking supplies
- Retrieving and storing evidence
- Documenting physical observations of an item of evidence
- Collecting evidence from an item and preserving it for testing
- Preparing samples for testing
- Running an instrument
- Reviewing instrument data
- Typing reports
- Performing a technical review of another scientist's work
- Participating in a pre-trial meeting
- Providing testimony
- Receiving or providing training
- Receiving continuing education
- Reading scientific literature or presenting a summary of the content of a scientific journal
- Participating in a proficiency test (depending on the discipline – annually or every 6 months)
- Performing a protocol review
- Participating in a self-assessment (audit of procedures/case reviews)
- Attending a staff meeting

While the above tasks are self-explanatory and observable, there are others that have more to do with the overarching **culture** and **language** of the crime laboratory system. But before we go, take note of the list again. Did you have a hunch that the tasks displayed above (gray text) are associated with quality assurance? In fact, while the highlighted tasks may be more obviously associated, one could argue that all of the tasks listed above are in some way tied to quality assurance.

The Quality System

The crime laboratory culture is established by the underlying quality system that is in place.

The quality system covers the laboratory goals & objectives, organization & management, personnel, facilities, evidence control, validation, analytical procedures, equipment calibration & maintenance, reports, review, proficiency testing, corrective action, audits, and safety.

The forensic scientist interacts and participates in this quality system in many ways. Here are a few:

Conduct – behaves according to the organization's code of conduct (professional behavior).

Hygiene & Dress Code – adheres to laboratory issued dress code and expectations for cleanliness and professional appearance.

Heads up! Dress codes can vary based on the task (i.e., court testimony vs. laboratory testing) and are often required for safety reasons (i.e., closed-toe functional shoes vs. sandals), to maintain the integrity of the evidence and the testing (i.e., lab coats, scrubs, gloves, etc.) or due to the public-facing nature of the role (e.g., presentations, courtrooms, professional meetings). Unlike the TV version of CSI, no one shows up to the lab wearing the sunglasses, sexy tank top and high heels combo.

Reality check #2

High heels are not comfortable. Rock the sneakers!

Procedures, Protocols, Validation – receives training, gains competency and remains proficient in adherence to procedures and protocols; may become involved in performing validation studies to effect procedural or protocol revisions.

Evidence Handling, Chain of Custody, Documentation – follows appropriate procedures to mitigate damage or loss, maintains complete and accurate records of when the evidence is moved or changes hands, and accurately and thoroughly records their examinations and findings.

Technical Review – performs secondary analysis or review of other forensic scientist's cases for the testing authorized to perform. Receives technical reviews of work by other forensic scientists proficient and authorized to review the same testing types.

Proficiency, Audits, Corrective Action – participates in regular testing to ensure the proper application of protocols and procedures; applies critical thinking skills to determine the root cause(s) of unintended or undesirable event(s) and recommends corrective measures.

Law and Legal Procedures – understands the basics of the judicial system and has been trained to present credentials, training, methods, and scientific findings in a court of law in an unbiased manner.

Communication

Human behavior – the way people act and react in response to a situation within a given environment, often ingrained in existing culture and communication patterns (i.e., boss-to-peer or peer-to-peer interactions). In order to be effective in a socio-technical environment one must understand how behavior impacts the trajectory of decisions. This nearly undetectable behavior can be the root cause of many unintended, undesirable and often damaging events.

Forensic scientists interact with many individuals throughout their day. This can be a supervisor assigning a case, a technician providing supplies or running instrumentation, attorneys inquiring about results, administrative staff assisting with report distribution, or a detective checking on the status of a case. Each of these instances presents an opportunity to communicate effectively. The following are just a few examples of how the scientist's **communication** skills are put to use during these interactions:

Instruction & Feedback – working in a laboratory may involve sharing instrument runs across multiple scientists' case samples (i.e., batching) or utilizing a technician's skills to perform a portion of the work. Whether it is a team-based approach or a case-to-case interaction with your supervisor/peer, this job requires an individual to:

✓ deliver clear instructions as to how work is to be performed,

✓ provide a thorough explanation when something does not go accordingly, and

✓ receive questions, suggestions, and feedback in a calm and non-confrontational manner.

Pre-trial & Testimony – present information in a clear and professional manner. Explain the purpose of the testing performed as well as any limitations of the test.

Human Factors – understand the human factors to improve reliability and reduce the opportunity for error or bias. Bring forth observations/awareness of potential gaps in knowledge, procedure, or training etc., discrepancies, errors, and misconduct.

Reality check #3   This isn't the wild west – there are no lone cowboys or cowgirls. The Forensic Scientist does not operate independently of everyone else.

While there are aspects of the daily work of a forensic scientist that are independent you must not forget the bigger picture. The scientist is part of the crime lab system, reliant upon the training program, technical procedures, equipment, facilities, and support of peers and managers to be successful. From gathering supplies to creating charts for courtroom testimony, the scientist must engage with many individuals throughout the day.

TIP: In today's technology driven world we develop communication short-cuts motivated by a sense of efficiency and/or comfort. Do not miss the opportunities for face-to-face human contact. Make it a personal goal to understand your communication tendencies and develop your communication style. Be observant to the communication styles of the individuals you interact with and seek to find ways to connect more often.

One final comment about communication... The ability to create an environment for connection and open dialogue is a valuable skill. The forensic scientists' casework (e.g., notes, sketches, test results, conclusions) is subjected to peer review by another qualified scientist. In forensics you do well if you can accept criticism and accountability for human error with humility and openness. The goal is NOT to retain an unbruised ego but rather to do the best work possible remembering it is the wellbeing of the community and the victims that you aid in your service.

It is up to each individual to determine how satisfying and effective their interactions will be. Ultimately, it is a choice – it is an attitude thing!

Through accreditation, standards, and a robust quality control system, everyone plays a role in ensuring the work that is completed is reliable. Aside from the work that happens at the scientist's laboratory bench, there is much more going on behind the doors of the crime laboratory than meets the eye (or the CSI television viewers' eyes).

In the next chapter we will zoom out, away from the scientists' lab bench and zoom in to the front steps of the crime lab building where we reveal the process you will take to make your way toward earning access to a position in the crime laboratory. If you haven't figured it out yet, they don't just let anyone in!

What's it Going to Take?

They don't just let anyone in! Understanding the educational and employment requirements.

You are off to a great start. Like a true scientist, you have done your due diligence in collecting as much data as you can about yourself, the profession, the high-stakes nature of crime laboratory work, and your specific forensic field of interest.

The previous chapter explained scientific and ethical standards, quality assurance, and the laboratory work environment and its place in the judicial system. Here we will cover the hiring (e.g., background check, professionalism) and educational requirements.

We start by revealing the hiring practices of the forensic laboratory industry, keeping in mind the standards at play.

Hiring Practices

As part of any hiring process you can expect the agency to require a candidate to:

- ✓ meet a set of *scientific* and *ethical standards*
- ✓ possess an understanding of a laboratory *work environment* and the *judicial system*

✓   pass a rigorous *background check*
✓   exhibit and embody *professional behavior*
✓   possess the necessary *education, coursework,* and *experience*

Background Check

Take a look at this excerpt from a job posting for a forensic scientist position at a city crime laboratory. Notice the *type of information* that is checked during a background investigation:

Conditions of Employment: The City's Background Check Policy requires background checks to be conducted on final internal or external candidate(s) applying for any position with the City of XXXX.

The type of information that will be collected as part of a background check includes, but is not limited to *reference checks, social security verification, education verification, criminal conviction record check, and, if applicable, a credit history check, sex offender registry and motor vehicle records check.* Final candidates must pass a pre-employment *drug-screening test.*

Notice the additional measure taken to screen candidates for *activities that are not acceptable* for jobs handling items related to a crime scene or criminal investigation.

"Final candidates must pass a pre-employment *drug-screening test.*"

Behavior – Past and Present

While certain information may be protected under fair employment laws and/or privacy provisions, the crime laboratory will screen potential employees for undesirable activities and behaviors. Remember, you are applying for a job that involves the "public trust." Crime laboratories are looking for personal characteristics such as honesty, integrity, and scientific objectivity. It's called a *back*-ground check because it is a search of both current and ***past*** information in order to determine if negative patterns of behavior have been corrected or still exist.

The length of the search for historical behavior may depend upon the age of the candidate. It's completely reasonable to assume that the older a candidate is the more history they have created. Therefore, the background investigator may require more time to review and vet their history. History can mean any of the following:

Criminal history – review of felony/misdemeanor convictions

Credit history – review of financial responsibility: handling credit, paying bills, and managing debt

Personal history – review of identity, past residences, education, employment, marriage, etc.

Social media history – review of content of text messages, blogs, videos, photos, etc., demonstrating illegal, biased, or immoral conduct

Heads up! An applicant's **social media posts** can assist in determining fitness for employment. [9]

Driving record – review of driving history: past tickets and driving violations or offenses

Drug Testing (history of drug use)

The level of scrutiny may depend upon the laws operating in the state you are applying for a job in, or the type of drug use under examination. Drug testing is performed to detect current use while the background check looks for offenses related to use/possession. The focus of this background check varies from agency to agency and is related to certain types of illegal drug use or possession.

Fingerprints

In addition to the collection and review of a candidate's past information (as listed above), fingerprinting is another screening tool to verify identity and review criminal records. Candidates submit to a fingerprint collection that is either ink and roll imprints or electronic fingerprinting (live scan). The fingerprints are then sent to the FBI and state bureau of investigation to be searched in the criminal databases. Any time a person has their fingerprints taken they will appear on a fingerprint background check. Fingerprints can be taken for criminal reasons and for civil reasons such as applying for certain licenses or permits. The check will provide the following information: name, aliases, residence, citizenship, place of birth, date of birth, sex, race, height, weight, eye color, hair color, and social security number.

The Polygraph Examination (AKA: lie detector test)

The polygraph examination is meant to test for truthfulness and to reveal immoral and criminal behavior. The polygraph device documents change in respiration, perspiration, and heart rate when a person is untruthful. The interviewer will ask questions about topics such as falsifying information, concealment of information that would impact your chance at employment, past and current illegal drug usage, theft, and violent or deviant behavior. It's natural to be nervous during this test. Each agency has its own policies as to what type of *minor* offenses will determine *pass* vs. *fail*.

The polygraph examination, drug testing, and all other aspects of the background check are methods the crime laboratory uses to evaluate and choose which candidates to advance or remove from the hiring process. The goal: to employ individuals who embody trustworthiness, truthfulness, integrity, impartiality, and confidentiality.

Your Employer – the Crime Laboratory: Requirements in the Field

If the crime laboratory will be your employer, you do need to figure out what area of the crime lab you want to work in. This section will help you identify requirements that are unique to the forensic science discipline(s) you are most interested in.

To begin we will review the *qualifications* section of job descriptions to tease out the requirements. When viewing job descriptions of any kind you should pay attention to the following information and whether it is preferred/desired or required:

✓ role (entry level vs. senior level, singular vs. dual role)
✓ education
✓ experience

Not all forensic science professions *require* the same **minimum qualifications**. The skills, certification(s), experience, and education required to be considered for a position are routinely found within the *qualifications* section. To illustrate this point let's take a look at the following excerpts from actual job postings to learn more about the distinctions.

In this first example, notice the **educational requirements** as highlighted below in the forensic toxicology job positing. In this posting the employer requires what is presumed to be a degree with a concentration in forensic science and a minimum number of credit hours of coursework in chemistry. Keep in mind, qualifications may vary depending on the employer and the specific job duties involved.

Crime Lab Scientist (Toxicology)

—A Forensic Science degree, which included the successful completion of at least 40 semester (or 60 quarter) hours of CHEMISTRY coursework
—Successful completion is considered a "C" or better in coursework

The **degree type** and **individual coursework** required of a forensic toxicologist, as shown above is different from the degree and specific coursework required of a forensic DNA analyst as shown below.

DNA Forensic Scientist

—Bachelor's degree in a Biology, Chemistry, or Forensic-related area and have successfully completed coursework covering the following subject areas: Biochemistry, Genetics, Molecular Biology, and Statistics and/ or Population Genetics.

The coursework identified above as a requirement for the DNA forensic scientist (above) follows the current *standard* for laboratory personnel (education, training, and experience) found in the *Quality Assurance Standards for DNA testing Laboratories* developed by SWGDAM through the FBI.

Look closely at the job posting to identify the **type of role** being offered. Laboratories associated with police/sheriff's departments may recruit candidates for positions in forensic science disciplines with a dual purpose. Take for example this job posting for a fingerprint examiner position:

Criminalist – Latent Print Examiner

—Bachelor's degree in forensic science, Criminal Justice, Biology, or a related science; or any equivalent combination of relevant training, education, and experience
—Must complete required training and be certified by the State of XXXXX Criminal Justice Academy as a Class 1 or Class 3 Law Enforcement Officer
—Candidate selected for this position will be required to obtain certification as a latent print examiner through the International Association for Identification within two (2) years of becoming eligible to do so, and are required to maintain such certification as an ongoing condition of employment

Take note of the title in the above posting. If you perform an internet search for the job title criminalist you would find that the role specifies duties in a forensic discipline like in the example above (criminalist – latent prints). You will also find that the criminalist title is often associated with the forensic biology discipline (serology & DNA).

Positions may differ based on **levels of experience**. Take for example these job postings for positions as drug chemistry examiners. Notice the number of years of experience required for each and the specific type of chemistry (analytical):

Forensic Scientist 1 - Drug Chemistry

—Bachelor's degree with a major in chemistry, toxicology, pharmaceutical science with chemistry concentration (organic, inorganic, analytical or physical), forensic science (with concentration in chemistry or toxicology), biochemistry, pharmacology, or closely related curriculum from an appropriately accredited institution; or an equivalent combination of education and experience

Management preference: Management prefers 2 years of related experience

Forensic Chemist 2

—Must possess a bachelor's degree in chemistry, analytical chemistry, pharmaceutics, or biochemistry

—Must have a minimum of four years of progressive analytical chemistry experience

Wow! Eye opening, right?

As you observed in the previous exercise, there are minimum requirements that must be met in order to be eligible to apply for a career in a modern forensic science laboratory. These requirements can differ based on the position/level you are applying for. You may have also noticed that certain credentials/skills are *preferred* as opposed to *required*. This means that the crime laboratory would select one candidate over another if all other things were equal, one candidate met all of the preferred requirements in addition to the minimum requirements.

Now that you have examined a few crime laboratory job postings you will have an opportunity later to perform an independent search for current openings in your field of interest. You will then use this information to create a checklist of requirements you will need to begin your assessment of the schools and programs that offer forensic science degrees.

Before launching into a search for the right college, you will want to be certain you understand all the requirements. This goes beyond knowing the conditions to be eligible to apply for a crime lab job. Simply picking a college with a degree program in 'forensic science' does not guarantee you will have the opportunity to apply for a job when that diploma is earned. To possess the desired qualities and skill set for your field of interest you want to focus on what the school can offer (e.g., specialized forensic courses, laboratory skills, forensic equipment and tools, current research, collaboration with the forensic community, etc.) in addition to the degree type.

Education and Coursework

As instrumentation and techniques in the crime laboratory advance so does the need for enhancing and standardizing the forensic science practitioner's training, education, and professional development programs. While forensic science laboratories provide on the job training there are educational requirements and preferred skills that many lab directors seek in prospective applicants.

Here we explore the current status of knowledge needs in forensic science and how to make use of this information when considering where and how to invest in your higher education goals. Why? Because there are so many choices...Why?

Truth #1 Interest in the field of forensic science exploded as the media reported on its crime solving power and the entertainment industries popularized the scientists' part.

Truth #2 The increase in educational institutions offering degrees in forensic science is a result of the increase in the field's popularity.

Truth #3 The work of the forensic science professional is critical to the criminal justice system. Their ability to perform their analyses and explain it in court emphasizes the need for appropriate training and foundational knowledge.

Back to the all-important question: How can you be sure you

know what it's going to take before you select the institution where you will study? Or, similarly, if you've already begun your college education, how do you tailor your program to insure you have what you need come time for graduation?

History Lesson

In order to talk about *requirements* it's best that you know why and how these educational requirements came to be to understand the importance of scrutinizing the curriculum of prospective academic programs.

To ensure that these programs would adequately prepare students for careers in the crime lab, the U.S. Department of Justice was directed to support an effort to ensure that "the Nation has an adequate pool of trained forensic scientists." Back in the early 2000s under the administration of the National Institute of Justice (NIJ), the Technical Working Group on Education and Training in Forensic Science (TWGED) was born. Each member of TWGED was, at the time, involved in educating and/or training forensic scientists and they themselves were a part of academia, forensic science laboratories, professional forensic science organizations and the legal system. This group developed a guide [10] that set forth recommendations for **career qualifications**, **academic curricula**, and **training and continuing education** for practicing forensic scientists. In turn, the NIJ guide established a need for a mechanism to formally evaluate forensic college degree programs – enter the Forensic Science Education Programs Accreditation Commission (FEPAC). Taking the recommendations from the guide, FEPAC developed a set of robust educational standards* [11] and defined an assessment process through which to evaluate forensic science academic programs. The goal – to provide students with the technical skills and coursework required to support the workforce needs of the forensic science community.

Accredited **degree programs** versus *non-accredited* degree programs: Academic institutions can seek to accredit a forensic science undergraduate or graduate degree program by applying for accreditation through the FEPAC. For more information on the forensic science programs FEPAC accredits and the FEPAC standards, visit the FEPAC website: https://www.aafs.org/FEPAC

Where FEPAC develops standards specific to college and university forensic science degree curriculum (coursework and skills), the OSAC's also play a role in establishing best practices relevant to the forensic scientist's training inside the crime lab. The OSAC documents outline the learning objectives, instructor qualifications, student requirements, syllabus, performance goals, periodic assessments, supervised casework period, training program assessment mechanisms, competency testing and evaluation of program efficacy and relevance in a 4-year period.[12]

Why is this important to you??? Because now that you are in-the-know about the different types of forensic education programs out there you can make an informed decision about the type of program that will best prepare you to meet your career goals.

Gathering Your Requirements for the College Search (filtering preferences)

In this section you will gather your requirements. At this stage in the game, having defined your area of interest will allow you to gather the most relevant requirements and tailor your checklist to a forensic science discipline (i.e., digital evidence, toxicology, crime scene, etc.).

ACTIVITY: Requirements in the Field

Start by gathering *your* requirements from the *field*. Search the internet for current job postings from your field of interest. You can use a separate sheet of paper at first to get a general idea of the requirements that are consistent across all of the postings you review (degree type, specific coursework, certifications, experience). Then transfer this information onto the *Career Requirements Checklist*.

The *Career Requirements Checklist* is part of a package of companion documents. **To access this document, please purchase the *Forensic Science Career Evaluation Workbook.***

Typical requirements fall under the following categories:

✓ **Type of Degree*** – Bachelor of Science, Master of Science, Doctorate
✓ **Type of Major** (depending on your area of study)
✓ **Mandatory Coursework** – i.e., FBI QAS required degree/ courses for forensic DNA analysts
✓ **Mandatory number of credit hours** – i.e., employer prefers concentration in a specific subject matter and/or set number of credit hours in a specific subject area (e.g., 30 credit hours in chemistry)
✓ **Computer science/information systems coursework or experience** (depending on your area of study)
✓ **Specialized training/Certifications**
✓ **Prior experience**
✓ **Background/Personal History**

While not a requirement, **cost** can be a key factor in selecting a location to study. With that in mind, if you choose community college as a starting point then you can still use the criteria you establish in this section and the selection criteria in the following chapters to guide your coursework selections to prepare you for your move to a 4-year school.

TIP: Utilize the resources of the career counselors at your school. Get set up to be on the most appropriate track to maximize the number of credits for transfer. You don't want to be in your final semester of community college to find that some credits/ courses will not transfer.

***A note about advanced degrees** (MS or PhD): Simply because an institution offers a graduate or doctoral degree program does not mean that the dollar value is worth the extra schooling.

Crime laboratories must provide on-the-job training in order to meet the needs of their laboratory specific services, procedures, instrumentation, etc. To this end, you should assess the curriculum of a graduate program in the same way you are evaluating the undergraduate program:

✓ Does the program better prepare me for the crime laboratory?
✓ Does the program provide more relevant, specialized laboratory experience?
✓ Does the program provide increased exposure to a crime laboratory and practitioners?

A STEM (science, technology, engineering, math) based foundational education with hands-on laboratory training and access to laboratory internships where relevant forensic research and/or validation studies are being performed can showcase a higher degree of career readiness.

This is something to consider as you give much thought to the institutions you wish to apply to for your undergraduate degree.

Yes, I did say institutions (plural) because you may not get into your top pick.

TIP: For entry-level positions, crime labs often hire candidates who earn a BS degree in a physical or biological science or the combination of a MS in Forensic Science (with a BS in physical or biological science). [13]

This is a technical field – laboratory, instrument, and research skills are super important.

8

Where Should I Study?

Evaluate schools and degree programs

Selecting the right school for you and your desired program of study is important for a number of reasons. You want your studies to be as applicable to the career field as possible. Equally important, your goal should be to find an academic institution that positions you to transition into an operational crime laboratory. The closer the educational environment reflects the culture and expectations in the professional environment, the better the transition.

> **TIP:** First, focus your evaluation on the characteristics of a good academic program... second, focus on the facilities and resources of the school and its' partnerships.

The forensic science program curriculum should provide a sound background in the natural sciences, require advanced science classes, and specialized coursework (i.e., computer sciences and other forensic discipline-specific methods/techniques, relevant forensic topics, etc.).

Practical application through laboratory instruction should

train students in current forensic techniques. Coursework specific to forensic science professional practice topics such as courtroom testimony, quality assurance, and ethics should be a required part of the curriculum.

If you don't have a specific program or institution in mind, you may find it helpful to visit a website dedicated to providing information on forensic higher-education programs:

https://www.forensicscolleges.com/
or
https://www.aafs.org/FEPAC

Remember, the most important aspect of the academic experience is the ability of the institution to meet your educational needs in preparation for the job market. Your evaluation must focus on how well the educational program AND institution prepare you for the career.

To do this you will need to capture the benchmarks on which to base your evaluation. Thankfully, I have done this for you.

To access the *Academic Program Evaluation Checklist* you will need to purchase the companion booklet: *Forensic Science Career Evaluation Workbook*.

SUGGESTION: You will get the most current job specific requirements by repeating the activities in Chapter 7 (Activity: **Requirements in the Field**)

Forensic Science Degree Program Evaluation

STEP 1. Evaluate program criteria - Focus on programs resulting in a Bachelor of *Science* (if you are seeking an undergraduate degree) or a Master of *Science* (if seeking an advanced degree) rather than a degree in the Arts/Management/Business fields. The degree program should be categorized as a degree in forensic science or a physical or biological science with forensic application.

Other points of evaluation:

✓ Degree type

✓ Discipline-specific Coursework and Forensic Science Topic Areas

✓ Laboratory co-requisites

✓ Specialized Science Coursework

✓ Internship*, Independent Study**, Research***

To see a more complete list of evaluation criteria, refer to the *Academic Program Evaluation Checklist – Program section*. This checklist is part of the companion booklet: ***Forensic Science Career Evaluation Workbook***.

* A note about Internships: It is no mystery that STEM (science, technology, engineering, and math) majors require some of the most challenging coursework and that it takes stick-to-itiveness to stay in the game and finish. Research indicates students who have the opportunity to complete an internship related to their field of interest are much more likely to persevere to graduation in their STEM program in their chosen field (as opposed to changing programs or dropping-out).[14]

An academic institution offering forensic science degree programs may have an agreement between the institution and the crime laboratory to facilitate internship opportunities. If an agreement is in place, learn more. For example, if there are multiple applicants, how are internships awarded (e.g., priority awarded based on number of credits earned, year in current program, graduate students over undergraduates)? If crime laboratory internships are not available – what alternative internship sites are possible (e.g., medical examiner, crime scene unit, police department, private lab, hospital lab, or research lab)? Is the internship simply a rotation to expose the student to the different forensic science disciplines in a crime lab or does it offer a project-based study (the latter being preferred)?

** A note about Independent Study Topics: Most crime laboratories have minimal resources for large projects and are unable to provide students access to facilities and records within the lab without a proper background check and training in confidentiality laws. Aside from limiting access, the topic of study or the project the crime lab may offer can be dictated largely by the laboratory resources and needs. That said, you want your project to be as relevant as possible and ultimately train you in a transferable skill. When evaluating a school, learn how the student's project is selected for internship or independent study – i.e., Is it based on relevant forensic science community needs or current/emerging applications to the field of forensic science?

> **TIP:** Think outside the box. If a crime lab internship is not available or bench time in a lab is not feasible, offer to investigate recent quality events, perform a root cause analysis, recommend a corrective action plan. This is an excellent way to obtain invaluable critical thinking skills and a better understanding of the quality assurance process – a necessary aspect of the scientist's job and a much-overlooked skill.

***A note about Research: Finally, independent research is equally valuable, if not more valuable than an internship if you are learning to apply the concepts of basic laboratory skills and problem-solving techniques. Gaining hands-on experience with instrumentation used in the modern-day crime lab is where you can practice your technique and develop troubleshooting skills. That said, if an internship or study at a crime lab is not possible, don't disregard the option of performing research in your department's lab – some academic institutions have impressive research facilities with instrumentation donated from crime laboratories. Combining this with a professor who has previous experience in the forensic science field, along with a mentor in your specific field of interest will prove equally valuable.

To access the full list of internship criteria use the *Academic Program Evaluation Checklist* – *Internship section*. This checklist is part of the companion booklet: ***Forensic Science Career Evaluation Workbook***.

STEP 2. Compare Programs – Review the curriculum and require-ments of the program. Use the checklist to ensure you can identify that the program meets your requirements. Place a checkmark next to each of your requirements on the checklist once you've con-firmed that the program meets or offers it.

STEP 3. Narrow your list – At this point you may have had academic institutions drop off your list because they did not meet the require-ments of your *program* and *career goals*. If this is the case, make note of why they fell short. If you do not gain acceptance to your top-ranking schools, you may need to revisit others that initially fell off your list. In this case, you want to know what you might not be getting out of the lower ranking institution so you can plan to sup-plement for this if you apply and attend one of these institutions.

STEP 4. Consolidate your list – Remove schools that you are not considering applying to (i.e., do not offer your program of inter-est, missing required criteria, etc.). The list should contain only the schools that made your cut.

Once you've selected *programs* that meet your requirements, now you are ready to assess the *institutions* to help you narrow down your list a bit further.

Academic Institution Evaluation

Nowadays, institutions post tons of information on their websites to showcase their facilities and programs. This is great for you because it makes it easier and less costly to vet schools based on what you can learn from the comfort of your home. Remember, all the work you did evaluating the *degree programs* should have caused you to update the list of potential candidate schools filtering down to only the ones meeting your *program criteria*.

STEP 1. Evaluate academic institution criteria – Focus on criteria that can provide both the educational resources and preparedness to apply for a role in the crime laboratory. You want to investigate the following:

- ✓ Facilities
- ✓ Faculty with Forensic Experience
- ✓ Affiliations/Partnerships
- ✓ Research Opportunities
- ✓ Resources

To see the full list of this author's evaluation criteria, use the *Academic Program Evaluation Checklist* from the companion booklet: *Forensic Science Career Evaluation Workbook*.

STEP 2. Compare Institutions – Use the checklist to ensure you can identify that each institution meets your requirements. Place a checkmark next to the requirement on the checklist once you have confirmed that the facility meets or offers it.

Most institutions offer virtual tours and opportunities to e-meet current students and faculty for a Q&A session.

That said, it can be even more helpful to schedule an on-campus tour to see the facilities and student culture for yourself. Visiting campuses is an added expense, and this is where narrowing your list of prospective schools in advance of scheduling tours becomes important.

STEP 3. Schedule Tours

While on campus (or by phone/virtually):

- ✓ Visit the campus program site and check out the classroom and laboratories
- ✓ Meet with instructors and students for a Q & A
- ✓ Learn about research opportunities – crime lab or research labs in proximity to the campus
- ✓ Assess school's job preparedness and placement by interviewing students
- ✓ Investigate school resources & social wellness –

collaborative nature (high/low) and student support networks (abundant/non-existent)

STEP 4. Assess Student–School Fit
When considering where to attend school it is wise to contemplate how you perceive your individual student-school fit to be. For example, how do you respond to different social and study environments?

✓ *Social Integration and Belonging*[15]: How do you believe the size of the school (size of the campus), the size of the classrooms (number of students per class), the on-campus living conditions, and the extracurricular activity offerings will impact your ability to engage with other students/faculty and form social networks?

✓ *Academic Performance*[16]: How do you believe your study habits, the accessibility of academic support services (such as tutoring), the program size (number of students in your program competing for same faculty/classroom resources), and the competitive nature of the coursework will impact your ability to perform and succeed academically?

✓ *Location and Resources:* How do you believe the location of the institution (metropolitan vs. rural), the availability of hands-on learning experiences (partnerships with area labs), student research opportunities, and career planning resources will impact your ability to place yourself in a career after graduation?

A student's social, familial, and financial support systems as well as pre-entry preparedness, existing study habits, awareness of and a willingness to engage in advising, mentoring, and tutoring resources can make a difference in whether a student thrives or barely survives the higher education experience.

TIP: Define the type of conditions and types of support that are most likely to provide an environment for you to thrive in.

STEP 5. Consider Financial Obligation – what is your budget? Will you need assistance? Is community college* a good first step? It is important to factor in your financial resources and personal financial responsibility when making your school selection. You do not want to incur added stress due to inadequate resources to maintain enrollment over the term of your program, potentially risking dropping-out.

> ✓ Investigate tuition assistance programs
> ✓ Understand your personal financial responsibility
> ✓ Determine the overall cost to complete the degree program (use this as a metric when comparing schools with similar degree programs, facilities, and resources)

*If you choose to attend a 2-year community college as a first step, be sure to do your homework: Does the community college have a relationship with your 4-year schools to ensure all credits transfer?
Work with an academic counselor at your community college to **Chart Your Path** (see Chapter 10).

STEP 6. Application Time – apply to the institutions/programs meeting your criteria and needs.
Have a back-up plan if you are not accepted upon initial application. Or apply to additional high-ranking institutions/programs that did not meet all of your criteria. Just keep in mind that you will want to account for anything that is lacking in order to supplement the deficiency.

> **TIP:** Rank the institutions you've applied to. Include a reason for each ranking, e.g., What benefit does it have over the next highest-ranking school? Keep track of this. If you enroll in an institution/program that does not meet all of your requirements, you'll need to make note of those requirements that were not met during your assessment so you can advocate and plan for what you need.

A Note about Accredited Forensic Science Degree Programs: FEPAC accredited academic programs are required to display information on their institution's website about the *success rates* of their programs and achievements of their students. This is an excellent way to gauge student experience and opportunities after graduation. To further evaluate a FEPAC accredited program's academic success you can visit the FEPAC website to see a list of schools with a FEPAC accredited program (https://www.aafs.org/FEPAC) and navigate to the 'Student Achievements' section. The type of information you might find (student enrollment numbers, graduation rates, post-graduation vocations, internship locations, senior projects, etc.) can give you a picture of the size and health of the program, the relevancy of the curriculum to the field, and the department's connectedness to the forensic community.

If the post-graduation information is collected as part of a voluntary survey, it is often not representative of the full complement of response from all graduates from any single graduating class. Interviewing current students and graduates of the program firsthand can help to supplement this information.

For students who are already attending college and enrolled in a non-forensic related program or a forensic science program that is not a FEPAC accredited program you can use the *Career Requirements Checklist* from the previous chapter to help navigate you directly to the chapter titled: *Chart Your Path*.

9

Prepare to Launch

It's never too early to start pre-college career preparation

Filling out college admission applications does not have to be the starting point for your preparation to launch a career.

You can begin to prepare for your future in forensic science now, even while still in high school, and at the same time better position yourself to get academic institutions and employers to notice you.

This is also another opportunity to use this time to investigate your fit. If these activities are not of interest to you then you had best reassess your goals.

Career Related Activities

Here is a list of activities to consider becoming involved in now:

- ✓ Register for math and science classes
- ✓ Seek out classes with laboratory components
- ✓ Develop good organization and record keeping skills (laboratory and classroom notes)
- ✓ Take a public speaking class or find opportunities to

present information in front of a group (i.e., teach a summer camp class, read the Bible verse at your church service, audition for a part in the school play, etc.)

✓ Watch real (not T.V. generated) courtroom proceedings to learn more about the judicial system and understand how cases are presented

✓ Attend a summer pre-college program in forensic science

✓ This is a fantastic way to test the water to see if you really like forensic science and to figure out which area(s) you most gravitate toward. Here are a few offerings that pop up with a quick internet search. Perform your own search to locate programs near you. Most seminars of this type are hosted by an academic institution but a city or county police department operating a crime laboratory may offer a similar program. Here are a few examples:

- The Center for Forensic Science Research & Education
- https://www.cfsre.org/education/the-forensic-sciences-mentoring-institute
- National Student Leadership Conference
- https://www.nslcleaders.org/youth-leadership-programs/forensic-science/
- University of Delaware's Forensic Science Pre-College Summer Program https://www.pcs.udel.edu/forensic-science/
- New Jersey Institute of Technology: Forensic Science Initiative https://www.njit.edu/precollege/fsi-forensic-science-initiative

✓ Independent study: Access articles and webinars on forensic science topics offered by forensic science organizations and partners such as:

- Forensic Technology Center of Excellence (FTCOE)
- American Academy of Forensic Sciences (AAFS)
- Research Triangle Institute (RTI) International
- DOJ-BJA Sexual Assault Kit Initiative (SAKI) Academy

✓ Micro-credentials: Investigate opportunities to earn micro-credentials in forensic science topics such as:

- Introduction to forensic science
- Digital critical and fundamental skills
- Crime scene photography

TIP: Keep a record of your activities – including forensic educational activities and hands-on experience. Not only will these activities highlight your interest and display your initiative, but this record can serve as a starting point to building a professional résumé.

✓ Young Forensic Scientists: Forensic science professional organizations and suppliers of forensic equipment and testing materials are supporting the next generation of forensic scientists with all types of resources. Check them out to learn how they might benefit you.
- American Academy of Forensic Sciences (AAFS) – Young Forensic Scientists Forum
- American Academy of Forensic Sciences (AAFS) – Student Academy
- International Symposium on Human Identification (ISHI) – Student Resources
- Promega – Young Forensic Scientists Resource Center
- Qiagen – Young Investigator Award

Gather References

Do you have educational or professional mentors to bear witness to your character – your qualities and abilities? This could be a teacher, manager, coach, or other professional who knows you and has been involved in some way with your training, work, and/or development. Reach out to these individuals and share with them how they have impacted your life and your pursuit of your interests. Next, gauge their willingness to provide you with a testimonial. If you have a relationship with a faculty member associated with a crime lab – can that individual speak of your performance?

TIP: Use testimonials to help you articulate skills that you can showcase in a professional résumé.

Make Connections

It is never too early to begin connecting with individuals who will support you and help you get your footing in a career. Much like the people you turn to as a *reference,* networking is a purposeful step to initiate a professional relationship with others in your field of interest. These relationships can keep you up-to-date on current information in your field, share knowledge and skills, and offer advice.

- Investigate professional organizations offering student engagement opportunities
- Engage with Scientists & Teachers

Some ideas...

Volunteer in Math or Science – Do you have a high school forensic science, chemistry, biology, physics, or math teacher that needs an aide to help with laboratory set up? An excellent way to foster connection is by becoming involved in someone else's work. Forensic work plays a role in serving the community through crime prevention and justice. Start *serving now*! and build a skill at the same time.

Get scientists into the classroom – Do you have a local or regional laboratory in a STEM field that might welcome the opportunity to visit your school to talk about their job? As a professional, it's rewarding to share your work with others and come alongside those who are just starting out to provide mentorship.

TIP: Networking lets organizations know you are a job seeker. START EARLY – employers love initiative!

Showcase Your Skills

It is a good idea to showcase your skills to a prospective employer. Knowing what they are and articulating them can be tricky. Use the activity below to define skills unique to you.

ACTIVITY: Skill Set Self-Evaluation – inventory of soft vs. hard skills

Take an inventory of the skills you possess. List an activity you enjoy doing and try to break it down into steps (as shown in the first example below). An individual step in the process of an activity might reveal an individual skill. Referring to a list of named skills or traits might also help.

soft skills: time management, organization, communication, presentation, flexibility, creativity, critical thinking, problem solving, etc.

hard skills: accounting, data analysis, computer skills, editing, writing, analytical skills, bookkeeping, language proficiency (multilingual or bilingual), typing speed, operating machinery, software knowledge, etc.

Referring to job postings can also provide insight into skills the employer values. Here are a few examples:

- Ability to communicate effectively, both orally and in writing.
- Ability to work effectively with co-workers and the general public.
- Possess excellent prioritization, time management, and organization skills.
- Ability to maintain a high degree of confidentiality.
- Must be able to evaluate and interpret observations, maintain accurate records, and solve technical problems.

ACTIVITY: Skill Inventory

Sometimes it helps to define skills by actually doing the thing that displays your skill set. We will do this on paper.

STEP 1. Pick an *activity* that you take pride in doing, enjoy doing, spend the most time doing, or get the most positive feedback from doing and list this at the top of a new column.

STEP 2. List each step involved in your activity in the rows below, in order (step 1, step 2, etc.)

NOTICE: To help get you started I have pre-filled the first column with an example.

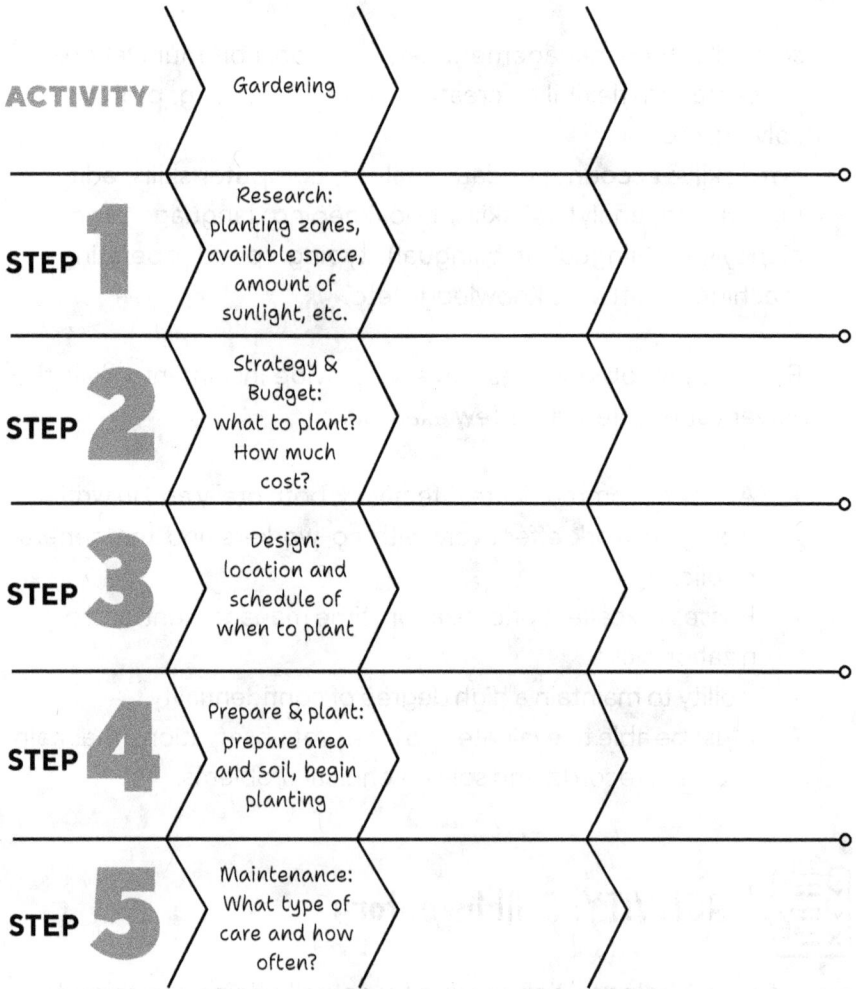

ACTIVITY	Gardening		
STEP 1	Research: planting zones, available space, amount of sunlight, etc.		
STEP 2	Strategy & Budget: what to plant? How much cost?		
STEP 3	Design: location and schedule of when to plant		
STEP 4	Prepare & plant: prepare area and soil, begin planting		
STEP 5	Maintenance: What type of care and how often?		

STEP 3. Review the steps in each of the activities you have plotted. In each step, look to see if there is an identifiable skill.
STEP 4. Circle or highlight the skill.
STEP 5. Consider using these named skills in the next section: Résumé Development.

Pull it all Together

Now that you have made this additional investment into your interests, let's put all of these activities and associations into an organized format to showcase education, skills, experience, etc.

Résumé Development

Most computer document creation software provides résumé templates. You can also search the internet for free templates. The template will guide you through the most common information you should include on your résumé. Once you have found a template, begin to populate the information for each section. Be sure to include the information that will display your interest in forensic science.

Focus on the following topics:

- ✓ highlight any coursework in forensic science, chemistry, biology, math, physics, or natural science that demonstrates forensic relevance
- ✓ highlight laboratory experience, including specific skills or techniques
- ✓ include any specialized training, skills, or experience in the field – did you have a class, job, volunteer, or shadowing experience that provided training or developed a specialized skill set?
    - note taking/documenting, writing, review, presentation, etc.
    - quality assurance or corrective action experience – problem solving, critical thinking, root cause analysis, etc.
    - jobs or coursework with relevant forensic lab work

- (training or experience utilizing instrumentation or technology)
- jobs or coursework with a law or criminal justice application
- photography, drawing, etc.

✓ include any independent/extra-curricular education activities
- certificates of completion for forensically relevant webinars, meeting attendance, asynchronous training modules
- list forensically relevant publications that you've read

✓ affiliations – list any professional organizations you belong to

✓ include soft vs. hard skills – what skills or abilities do you possess that fit well with this area of study/work?

Reminder: Use your testimonials and the list of skills you gathered from the self-evaluation activity (Chapter 3) above if you need help identifying them. Revisit your personality assessments (Chapter 4) and your Holland Code Career test report (Chapter 5). Still having trouble? Try the skill set self-evaluation above.

Once you have completed your first draft of your résumé, seek out someone to review, edit, and advise you as you develop your résumé. Choose a 'someone' who knows you and your skills so you can get the most personal and tailored advice. If you have the connections, you may even wish to pass a copy to a professional in your field of interest* or your guidance counselor for additional insight. Gain the wisdom of others. A fresh set of eyes can reveal an opportunity that you may not have otherwise seen. Especially if you have been staring at the document for hours!

Finally, remember that résumé development is 'development' because it is an ongoing activity. You never stop learning and building your capacity. You will want to review your résumé from time to time to ensure it is current, adding any new skills, training, education, and experience gained as the months and years go by.

TIP: Résumé development is an ongoing activity. You never stop learning and building capacity.

*Need a professional in the field to help you evaluate your résumé? Don't be shy. Find my contact information in the More About the Author section.

Chart Your Path

Get the most out of your education: discover and maximize school offerings and begin the career alignment process

Road Map to the Finish Line

Now that you have been accepted into an academic institution, it is important to start planning as early as possible for the job market by maximizing all that your academic institution offers while you are attending classes.

You may feel as if you are done and DONE with this process. But please don't bail out now. This is the most important part, seeing your goal through to reality. That means planning. You need to plan ahead to maximize your educational program's resources, allow time to develop new skills, and schedule opportunities to gain exposure to your field.

Notice the key words in the previous sentence: plan *ahead*, allow time, and schedule. The precursor to planning is **prioritization**. I can't emphasize this enough. **Decide the order and then plan**. There will be events that need to be sequenced in order to reach the goal. Be aware of any prerequisites, application processes, or letters of reference requirements to factor into your timeline and necessary resources to get to the end goal. For example, if you are interested in securing an internship at a specific location, consider the following:

Does the site have an application process?

Is there a timeframe when applications are accepted?

Are transcripts or letters of reference required?

Must you be enrolled in a specific higher ed program to be eligible to apply?

Are subject-specific coursework or a number of credit hours a requirement?

Will a background check be performed?

As this example demonstrates, you must understand what is required in order to plan. The biggest part of planning is prioritizing appropriately. The secret: always allow yourself plenty of time.

The Road Map is a checklist of all the requirements and goals you have collected along the way:

- ✓ academic requirements
- ✓ laboratory experience
- ✓ specialized training – instrumentation & hardware/software
- ✓ public speaking experience – presenting scientific data
- ✓ exposure & experience in the field
  - tours/shadowing (see a model: development of the Forensic Investigative Curriculum (FIC) School – Educational Outreach Initiative – NYSP CLS)[17]
  - volunteering
  - employment relevant to laboratory work
- ✓ internship/research project in relevant forensic science topics
- ✓ hard & soft-skill development
- ✓ professional associations & relationships

Having previously been involved in interviewing and hiring candidates for crime lab technician and forensic scientist positions,

I can't stress enough the importance of acquiring a strong knowledge base in science and math, an understanding of the principles foundational to forensic testing methods, and a familiarity with forensic technology/instrumentation as well as an awareness of its limitations.

> **TIP:** A bit of advice – keep in mind the career environment you will be entering. It's a laboratory. It will serve you well to not blow off your lab classes but rather, use that time to develop career-ready skills.

Be savvy, learn and practice lab etiquette – professional behavior in a lab means taking precautions to maintain a safe, clean, and respectful environment. Make it your practice to minimize distractions or disturbances to yourself and others by scheduling your time, organizing your equipment and supplies, and maintaining a clean work area.

Ensure that you seize every opportunity to build solid laboratory foundational skills. Develop a clean technique to preserve evidence and minimize routes of contamination, practice and build pipetting skills, learn how to perform serial dilutions, and know the chemistry behind solution preparation. Read and understand your protocols.

Increase problem solving abilities, seize opportunities to use critical thinking skills. Understand why and how mistakes have been made and learn how to develop a preventative mindset.

Plan now to not only maximize your school's offerings but also their resources. You're paying for it. So use it! You wouldn't save up for a new laptop computer and then go out and grab the first one you see simply because it comes in your favorite color, would you? You would do your research – desirable features (touchscreen, weight/size – travel friendly, integrated software, etc.), accessories needed (extra battery/cords, headphones, etc.), and cost (how much time you'll need to earn the funds). Your education is a major investment in both time and money to build the foundation that is needed to launch you into your career. Start the career alignment process early.

To do this we will combine information from your ***Academic Program Evaluation Checklist*** (program requirements) and

Career Requirements Checklist along with *field readiness goals* to chart a road map to your career goal. This document will serve as a planning guide and progress tracker. You will gather the requirements necessary to reach your goal by plotting and scheduling out each step. As you complete each milestone you will track your accomplishments. Unforeseen requirements or challenges can be accommodated into the plan more easily when you have the overall process mapped.

The *Chart Your Path Road Map Checklist* is part of a package of companion documents. To access, purchase the *Forensic Science Career Evaluation Workbook*.

Let this chart be a source of encouragement to you. Start planning now!

You CAN do this!

A note about this journey I have taken you on:

I can tell you that this decision is the most crucial decision of your life – where you choose to go to school and what you choose to study. But the reality is, it is not.

Learning about who you are and what you have been *purposed for* is far more important.

Adjusting your attitude toward 'work' is what transforms work from burden to purpose. The way to purpose begins at the foundation of your being. What do you value? How can you serve?

Anything that is worth doing will take faith and hard work.

Yet, what is most important? Movement. Transitioning from student of self (what you now know about yourself and your purpose) to student of action.

You were created to participate and contribute. A wise pastor says it best (and I paraphrase): *Your life is not yours to keep to yourself, but rather you are to use your abilities to help those around you.*[18]

Get out there and live out your purpose!

ENDNOTES

1. Minnesota Bureau of Criminal Apprehension, *"Guidelines for Obtaining Known handwriting Standards"*, p.6. Available online. Archived 2010-02-02 at the *Wayback Machine*

2. CleanPNG, *Card Reader Machine for Data Input*, https://www.cleanpng.com/png-scantron-corporation-test-education-school-teachin-3688942/download-png.html

3. Wikipedia contributors, "Analytical psychology," *Wikipedia, The Free Encyclopedia,* https://en.wikipedia.org/w/index.php?title=Analytical_psychology&oldid=1241950188. Carl Gustav Jung, *Psychological Types: Collected Works of C. G. Jung* (Princeton University Press: Volume 6, 1971).

4. Wikipedia contributors, "Carl Jung," *Wikipedia, The Free Encyclopedia,* https://en.wikipedia.org/w/index.php?title=Carl_Jung&oldid=1242672959

5. Truity, *Using The Holland Codes for Career Planning*, https://www.truity.com/blog/page/using-holland-codes-career-planning

6. Stephen P. Robbins and Timothy A. Judge, *Essentials of Organizational Behavior:* Job Fit activity: Job Characteristics Model (JCM), (Pearson, 14th Edition, 2018), 121-122.

7. Linzi Wilson-Wilde, "The international development of forensic science standards — A review," *Forensic Science International*, July 2018.

8. National Research Council of the National Academies, Committee on Identifying the Needs of the Forensic Sciences Community, *Strengthening Forensic Science in the United States: A Path*

Forward, Washington, DC: The National Academies Press, 2009 *Strengthening Forensic Science in the United States: A Path Forward*

9.   Matthew O'Dean, Ph.D., *Conducting Social Media Checks of Police Applicants,* LinkedIn.com, https://www.linkedin.com/pulse/conduct-ing-social-media-checks-police-applicants-o-deane-ph-d-/

10. National Institute of Justice (NIJ), *Education and Training in Forensic Science: A Guide for Forensic Science Laboratories, Educational Institutions, and Students, NCJ # 203099* (National Institute of Justice, June 2004).

11. American Academy of Forensic Sciences (AAFS), https://www.aafs. org/FEPAC, *FEPAC Accreditation Standards.*

12. National Institute of Standards and Technology (NIST): US Department of Commerce, *The Organization of Scientific Area Committees (OSAC) for Forensic Science, OSAC Standards and Guidelines, updated February 2019,* https://www.nist.gov/organization-scientif-ic-area-committees-forensic-science/osac-standards-and-guide-lines

13. Mark Marohl, Grace Jensen, and Heather Barkholtz, "What is the Preferred Educational Background of Forensic Scientists?" *Journal of Analytical Toxicology,* Volume 47, Issue 3, April 2023, Pages 299–304, https://doi.org/10.1093/jat/bkac077

14. Stevens Institute of Technology, Media Releases (posted January 10, 2023), *Why don't students stick with STEM degrees?* https://www. stevens.edu/news/why-dont-students-stick-with-stem-degrees

15. C. Calle Müller, M. Kayyali, and M. ElZomor, (2023, June), *Board 139: Factors Affecting Enrollment, Retention, and Attrition of STEM Undergraduates at a Minority Serving Institution,* Paper presented at 2023 ASEE Annual Conference & Exposition, Baltimore, Maryland, 10.18260/1-2-42461, https://peer.asee.org/board-139-factors-af-fecting-enrollment-retention-and-attrition-of-stem-undergradu-ates-at-a-minority-serving-institution

16. Chunmei Chen, Fei Bian, Yujie Zhu, "The relationship between social support and academic engagement among university students: the chain mediating effects of life satisfaction and academic motivation," *BMC Public Health,* 2023;23(1):2368, Published 29 Nov 2023, doi:10.1186/s12889-023-17301-3, https://www.ncbi.nlm.nih.gov/pmc/articles/PMC10688496/

17. Ray Wickenheiser, Amanda Cadau, Claire Muro, Samantha Whit-field, Carrie McGinnis, Lola Murray, Melissa France, Lyn Niles, Donna Barron, Lori Valentin, "The forensic educational outreach initiative – Bridging the gap between education and workplace," *Forensic Science International: Synergy, Volume 8, 2024,* 100448, ISSN 2589-871X, https://doi.org/10.1016/j.fsisyn.2023.100448, or https://www.sciencedirect.com/science/article/pii/S2589871X23001353

18. Rick Warren, "God Gives You Abilities to Help Others," *Pastor Rick's Daily Hope,* August 24, 2022, from "Live Your Calling: What On Earth Am I Here For?" https://pastorrick.com/god-gives-you-abili-ties-to-help-others/

BIBLIOGRAPHY

A & E Cold Case Files (TV Series). Season 5 Episode 11 (aired June 3, 2006). A Killer's Skin/Where's Peggy. https://www.imdb.com/title/tt6291578/fullcredits?ref_=ttfc_ql_1

American Academy of Forensic Sciences (AAFS). https://www.aafs.org/ FEPAC. FEPAC Accreditation Standards.

Chen, Chunmei, Fei Bian, and Yujie Zhu. "The relationship between social support and academic engagement among university students: the chain mediating effects of life satisfaction and academic motivation." BMC Public Health. 2023;23(1):2368. Published 29 Nov 2023. doi:10.1186/s12889-023-17301-3. https://www.ncbi.nlm.nih.gov/pmc/articles/PMC10688496/

CleanPNG. Card Reader Machine for Data Input. https://www.cleanpng.com/png-scantron-corporation-test-education-school-teachin-3688942/download-png.html

Jung, Carl Gustav. Psychological Types: Collected Works of C. G. Jung. (Princeton University Press: Volume 6, 1971).

Marohl, Mark, Grace Jensen, and Heather Barkholtz. "What is the Preferred Educational Background of Forensic Scientists?" Journal of Analytical Toxicology, Volume 47, Issue 3, April 2023. Pages 299–304. https://doi.org/10.1093/jat/bkac077

Minnesota Bureau of Criminal Apprehension, "Guidelines for Obtaining Known handwriting Standards", p.6. Available online. Archived 2010-02-02 at the Wayback Machine

Müller, C. Calle, M. Kayyali, & M. ElZomor. (2023, June). Board 139: Factors Affecting Enrollment, Retention, and Attrition of STEM Undergraduates at a Minority Serving Institution. Paper presented at 2023 ASEE Annual Conference & Exposition. Baltimore, Maryland. 10.18260/1-2-42461. https://peer.asee.org/board-139-factors-affecting-enrollment-retention-and-attrition-of-stem-undergraduates-at-a-minority-serving-institution

National Institute of Justice (NIJ). Education and Training in Forensic Science: A Guide for Forensic Science Laboratories, Educational Institutions, and Students, NCJ # 203099 (National Institute of Justice, June 2004).

National Institute of Standards and Technology (NIST): US Department of Commerce. The Organization of Scientific Area Committees (OSAC) for Forensic Science, OSAC Standards and Guidelines. Updated February 2019. https://www.nist.gov/organization-scientific-area-committees-forensic-science/osac-standards-and-guidelines

National Research Council of the National Academies, Committee on Identifying the Needs of the Forensic Sciences Community. Strengthening Forensic Science in the United States: A Path Forward. Washington, DC: The National Academies Press, 2009. https://www.ojp.gov/pdffiles1/nij/grants/228091.pdf

O'Dean, Ph.D., Matthew. Conducting Social Media Checks of Police Applicants. LinkedIn.com. https://www.linkedin.com/pulse/conducting-social-media-checks-police-applicants-o-deane-ph-d-/

Robbins, Stephen P., and Timothy A. Judge. Essentials of Organizational Behavior: Job Fit activity: Job Characteristics Model (JCM). (Pearson, 14th Edition, 2018).

Stevens Institute of Technology. Media Releases (posted January 10, 2023). Why don't students stick with STEM degrees? https://www.stevens.edu/news/why-dont-students-stick-with-stem-degrees

Truity. Using The Holland Codes for Career Planning. https://www.truity.com/blog/page/using-holland-codes-career-planning

Warren, Rick. "God Gives You Abilities to Help Others." Pastor Rick's Daily Hope. August 24, 2022. From "Live Your Calling: What On Earth Am I Here For?" https://pastorrick.com/god-gives-you-abilities-to-help-others/

Wickenheiser, Ray, Amanda Cadau, Claire Muro, Samantha Whitfield, Carrie McGinnis, Lola Murray, Melissa France, Lyn Niles, Donna Barron, Lori Valentin. "The forensic educational outreach initiative – Bridging the gap between education and workplace." Forensic Science International: Synergy, Volume 8, 2024. 100448, ISSN 2589-871X. https://doi.org/10.1016/j.fsisyn.2023.100448, or https://www.sciencedirect.com/science/article/pii/S2589871X23001353

Wikipedia contributors. "Analytical psychology." Wikipedia, The Free Encyclopedia. https://en.wikipedia.org/w/index.php?title=Analytical_psychology&oldid=1241950188

Wikipedia contributors. "Carl Jung." Wikipedia, The Free Encyclopedia. https://en.wikipedia.org/w/index.php?title=Carl_Jung&oldid=1242672959

Wilson-Wilde, Linzi. "The international development of forensic science standards — A review." Forensic Science International. July 2018. National Institute of Forensic Science, Australia New Zealand Policing Advisory Agency, L6, T3, WTC, 637 Flinders Street, Melbourne, VIC 3008, Australia

ILLUSTRATIONS

Cover and illustrations: book cover inspiration; black and white sketches on cover and in Chapter 2; fishing girl and sea turtle in Chapter 3. An enormous thank you to my stepdaughter, the beautiful and talented artist **Caroline Kelly Donnelly**.

Graphics: Evidence Pathway, Syncing Up Your Personality Traits, Job Characteristics. Thank you to **Randi Harvey**, RH Publishing, for creating fun graphics for these activities.

Scientific method graphic (Intro section), check yourself–step 5 interests and values chart (chapter 3) by **Ashley Halsey.**

Icons courtsey of The Noun Project:
Magnifying Glass by B. Agustín Amenábar Larraín from Noun Project (CC BY 3.0)
caution sign by Tinashe Mugayi from Noun Project (CC BY 3.0)
Justice by nasril from Noun Project (CC BY 3.0)
Card by Roat Studio from Noun Project (CC BY 3.0)
person by Rusma Ratri Handini from Noun Project (CC BY 3.0)
Social Media by Imron Sadewo from Noun Project (CC BY 3.0)
Car by Nathaniel S. from Noun Project (CC BY 3.0)
attendant list by Candy Design from Noun Project (CC BY 3.0)
open book by Fitrapratama from Noun Project (CC BY 3.0)
report by uyun from Noun Project (CC BY 3.0)
hypothesis by syafii5758 from Noun Project (CC BY 3.0)
chat by Przemyslawk from Noun Project (CC BY 3.0)
bugs by QualityIcons from Noun Project (CC BY 3.0)
Hand by Lars Meiertoberens from Noun Project (CC BY 3.0)
Bug Search by Irfan ms from Noun Project (CC BY 3.0)

Sidebar and box art: Ihor Melnyk/istockphoto.com
Torn newspaper art: Kwangmoozaa/shutterstock.com

EDITING

To my editor **Laura G Johnson** of **Discover Your Voice.** Thank you for setting your eyes on this manuscript and for your expertise and insightful remarks. It has been such a pleasure working with you.

ACKNOWLEDGMENTS

During the initial drafting stages of the book outline and scope, I was fortunate to have the feedback of forensic experts and fellow consultants:

Lori Ana Valentin, Radiant Journey, LLC https://www.radiant-journey.org/ and

Ciaran Phillip, Sure Science Consulting, https://www.facebook.com/surescience/ https://www.linkedin.com/company/sure-science surescienceconsulting@gmail.com.

Thank you, Lori Ana and Ciaran, for your thoughtful remarks and business savvy. The structure and flow of the book would not be the same without your wise counsel.

A massive thank you to the following forensic science experts for their contributions of content review and revisions having made these sections (Forensic Science Professions – Chapter 2) as accurate and current as possible:

Crime Scene Search & Processing – **Christy Price, Kristine Woodhouse**

Forensic Toxicology – **Amanda Kogelschatz**

Forensic Biology (Serology & DNA Analysis) – **Meegan Fitzpatrick, Allyson Goble, Timothy Goble, Elizabeth Staude**

Forensic Chemistry – **Samantha Whitfield**

Digital & Multimedia Forensics – **Christopher McNeil, Jason Armstrong**

Trace Evidence Analysis – **Bradley Brown**
Firearms Analysis – **Matthew Kurimsky**
Fingerprint Analysis – **Andrea Lester** (Image of coffee mug with prints – **Eric Smith**)
Questioned Document Analysis – **Jill Dooley**

To my nephew and expert beta-tester, **Jonas Ethan Hekel**. The activity sections in this book are much improved as a result of your testing, observations, timing & record-keeping skills – a scientist at heart! Nice work, Partner!

To my brother, **Bryan Zevotek**, for sharing your wisdom on *connection through storytelling* as well as your constant supply of recommendations for good reads to fuel our shared constant-learner appetites. You can find Bryan at Classkickers.com or https://kicksomeclass.com/.

Much gratitude to my husband, **John** (AKA, The Idea Bouncer). You are my biggest fan and supplier of daily support, encouragement, and love. I am blessed!

To my parents, **Carol (Niebling) and Walter Zevotek**, thank you for love – a home, family, experiences, and adventures that shaped my values to this day.

MORE ABOUT THE AUTHOR

Nicole Zevotek resides in Charlotte, NC, with her husband John and his three beautiful daughters. They enjoy exploring nature and sharing the day's highlights at the dinner table or their favorite sushi restaurant. As a scientist, Nicole has a love for cooking and baking – articles from America's Test Kitchen are a favorite! Quality time with her husband looks like home renovation, gardening, hiking, bike riding to find a fun lunch spot and personal fitness. Her faithful dog Denali is the most enthusiastic companion to hit the trails with. She covets her morning runs and spiritual study time as regular practice to help her prepare for the day. You can find Nicole at linkedin.com/in/nicole-zevotek.

www.ingramcontent.com/pod-product-compliance
Lightning Source LLC
Chambersburg PA
CBHW052136270326
41930CB00012B/2904